U0192878

FANGCANG SHELTER HOSPITALS FOR COVID-19

CONSTRUCTION AND OPERATION MANUAL

方舱庇护医院

建 设 运 营 手 册

阎　志　主编

Yan Zhi　Editor-in-Chief

中国协和医科大学出版社

图书在版编目（CIP）数据

方舱庇护医院建设运营手册 / 阎志主编. —北京：中国协和医科大学出版社，2020.6

ISBN 978-7-5679-1539-8

Ⅰ. ①方… Ⅱ. ①阎… Ⅲ. ①传染病医院－建筑设计－手册 Ⅳ. ①TU246.1-62

中国版本图书馆CIP数据核字（2020）第083333号

方舱庇护医院建设运营手册

主　　编：阎　志
责任编辑：顾良军

出版发行：中国协和医科大学出版社
（北京市东城区东单三条9号　邮编100730　电话010-65260431）
网　　址：www.pumcp.com
经　　销：新华书店总店北京发行所
印　　刷：北京新华印刷有限公司

开　　本：889×1194　1/32
印　　张：5.75
字　　数：135千字
版　　次：2020年6月第1版
印　　次：2020年6月第1次印刷
定　　价：82.00元

ISBN 978-7-5679-1539-8

总 目 录

FANGCANG SHELTER HOSPITALS FOR COVID-19

CONSTRUCTION AND OPERATION MANUAL

方舱庇护医院

建 设 运 营 手 册

阎 志 主编

主　　编：阎　志

编写人员：阎　志　张　亮　杜书伟　方　黎

潘子晶　程龙华

英文翻译：阎　格

英文审校：王斯达

版式设计：黄　萱　宋　杰　叶芹云

前　言

新型冠状病毒（SARS-CoV-2）是一种新出现的病原体，具有传染性强、传播速度快等特点，主要通过飞沫和密切接触等方式，在感染者与被感染者之间发生传播。中国和世界其他国家及地区均出现了2019冠状病毒疾病（COVID-19，以下简称新冠肺炎）疫情。

为应对新冠肺炎疫情的暴发，方舱庇护医院作为一种新型公共卫生理念，于2020年2月由王辰院士在中国湖北省武汉市首次提出。在医疗资源有限的情况下，为解决数量急剧增加的新冠肺炎确诊患者得不到及时医疗救助的问题，王辰院士建议可将会展中心、体育馆等大型场所用最短的时间、以最小的成本改造成为可收治大量轻症和普通患者的方舱庇护医院。方舱庇护医院具有隔离、分诊、提供基本医疗照护、密切监测和快速转诊、提供基本生活和社会活动五个主要功能，在最大限度上收治新冠肺炎轻型和普通型患者，发挥迅速切断感染源、有效扩充救治床位的重要作用。

本手册由卓尔公益基金会组织参与方舱庇护医院建设、保障服务相关人员，对方舱庇护医院设计、改造、运营管理的实践进行整理归纳，从方舱庇护医院的提出、设计、改造、运营、保障五个方面总结经验，供全球各地区疫情防控工作同行参考。

阎　志
卓尔公益基金会创始人
2020年4月

目　录

第一章　提出建设方舱之策

1.1　方舱庇护医院建设背景

2020年1月23日，为阻断新型冠状病毒的传播，坐拥千万人口的超大城市武汉市封城。由于新冠肺炎确诊人数快速增长，医疗资源不堪重负，出现“一床难求”的局面，大量确诊患者无法入院只能居家隔离。2月1日，中国工程院副院长、中国医学科学院院长、呼吸病学与危重症医学专家王辰院士支援武汉疫情一线，在充分调研后，为落实中央“应收尽收、应治尽治”的方略，提出尽快建成方舱庇护医院的构想。2月3日，武汉市随即征用武汉客厅中国文化博览中心、武汉国际会展中心和洪山体育馆，搭建武汉市首批方舱庇护医院。

方舱庇护医院的建设主要利用既有的建筑或资源，在最短的时间内、以最小的成本建成，从而在最大程度上收治新冠肺炎轻型和普通型患者，既能让患者得到基本的医疗照护、向定点医院转诊及生活保障，同时还能有效地控制传染源、切断新冠肺炎病毒的传播途径，避免疫情加剧，实现提高治愈率、降低病死率的目标。方舱庇护医院的建设，虽然不是“至善之策”，却是可取之策、现实之策。在中国这次疫情防控工作中，方舱庇护医院发挥了重要的作用，也为其他国家应对此次疫情，迅速扩充医疗资源提供了新的思路、创造了新的模式。

全球2019冠状病毒疾病确诊病例数量的不断攀升，造成了各国医疗资源尤其是收治确诊病患的病床数量严重不足，导致相当比例的确诊患者无法及时收入医院以进行隔离治疗。居家隔离会使患者家庭成员处于危险之中，造成交叉感染，导致疫情进一步扩散、传播和蔓延。方舱庇护医院建立后，可采用科学的分类救治措施，将重症与危重症患者收治在定点医院，而方舱庇护医院则集中收治大量轻型和普通型患者，提高医疗资源的使用效率。

1.2 方舱庇护医院的定义

中国的方舱庇护医院是一种大型临时医院，通过将会展中心和体育场馆等现有大型场所改建为医疗设施而迅速建成，其任务是将大批新冠肺炎轻型和普通型患者与家人、社区隔离开，同时提供医疗照护、疾病监测和转诊、生活和社交空间。

1.3 方舱庇护医院的特点

全球顶级医学期刊《柳叶刀》(*The Lancet*)于4月2日发表了一篇聚焦中国建设和启用方舱庇护医院的文章《方舱庇护医院：应对公共卫生紧急状况的新理念》，该文章由王辰院士团队与德国海德堡大学全球公共卫生研究所等团队共同完成。文中列举了方舱医院的三大特点：建设快，规模大，成本低。这三大特点使其能很好应对公共卫生紧急情况。

方舱庇护医院主要收治那些未能入院的患者，既避免对家人造成传染，又能够使其得到及时医疗救治。方舱收治的全部是新冠核酸检测为阳性的患者，且通过检测，排除流感的感染因素，再通过要求收治患者佩戴口罩等措施，所以在方舱医院的收治患者之间基本不存在交叉传染的问题。

1.4　方舱庇护医院的功能

方舱庇护医院具有五大基本功能：

（1）隔离。收治轻度至中度症状的患者，其隔离效果优于家庭隔离。

（2）分诊。为确诊的新冠患者提供了一个战略性的分诊功能，轻度至中度COVID-19患者在方舱医院隔离治疗，而重症与危重症COVID-19患者在传统医院接受治疗，有效释放传统医院的压力。

（3）提供基本医疗照护。包括抗病毒、退热和抗生素治疗、氧气支持和静脉输液，以及心理健康咨询。

（4）密切监测和迅速转诊。每天对患者的呼吸频率、体温、氧饱和度等多次测量，对符合一定临床标准的患者，迅速转入指定高级别医疗机构治疗，大大缩短了从出现严重症状到被送往更高级别医院的时间。

（5）提供基本的生活和社会活动。在方舱庇护医院这种轻型和普通型患者社区，通过医护人员与患者以及患者之间相互协同互助及社交活动，缓解患者由疾病与隔离带来的焦虑，促进康复。

第二章　方舱庇护医院工程设计

2.1　方舱庇护医院功能分区

方舱庇护医院应根据功能合理地分为“三区两通道”，按医患分离、洁污分离的流线组织交通，采用负压通风系统，并预留适度的患者活动空间。

“三区”为污染区、半污染区和清洁区。其中，污染区包括轻症患者接受诊疗的区域、病床区、观察救治室、处置室、污物间以及患者入院出院处理室；清洁区包括更衣室、配餐室、值班室及库房；半污染区指位于清洁区与污染区之间、有可能被患者血液、体液等污染病毒的区域，包括医务人员的办公室、治疗室、护士站、医疗器械等处理室、内走廊等。不同区域应设置明显标识或隔离带，可采用不同色彩的标识进行区分。“两通道”则是指医务人员通道与患者通道，两通道应该完全分开。

清洁区进出污染区，出入口处分别设置进入卫生通过室和返回卫生通过室。进入流程为：“一次更衣→二次更衣→缓冲间”，以供医护人员穿戴防护装备后，从清洁区进入到污染区。返回流程为：“缓冲间→脱防护服→缓冲间→脱隔离服淋浴→更衣”，之后从污染区返回清洁区，且返回卫生通过室应男女分设。

2.2　方舱庇护医院病床区设计

病床区应做好床位分区、男女分区，每区床位不宜大于42床，每个分区应有2个疏散出口，分区内任一点至分区疏散出口的距离不大于30米，分区之间应形成消防疏散通道，高大空间内分区间消防疏散通道宽度不宜小于4米。分区内通道及疏散通道地面应粘贴地面疏散指示标志。分区隔断材料应选用难燃材料或不燃材料，表面耐擦洗，高度不宜小于1.8米。床位的排列应保持合适的距离，利于医生看护和治疗，平行的两床净距离不宜小于1.2米，并设置床头柜。双排床位（床端）之间的通道净距离不宜小于1.4米；单排床时，床与对面墙体间通道净宽不宜小于1.1米。

2.3　方舱庇护医院厕所设计

病患和医护人员厕所须分开设置。病人使用临时厕所，临时厕所区域与病房区域之间设置专用通道，优先选用泡沫封堵型移动厕所，厕所数量按照男厕每20人/蹲位、女厕10人/蹲位配置，可依据病人实际需求适当增加，厕所位置应在方舱下风向并尽量远离餐饮区和供水点。病人使用临时厕所产生的生活污水与洗浴废水必须经过专用蓄水池消毒处理，严禁未经消毒处理或处理未达标的病区污水、医疗污水、病区污染物直接排放至市政污水管道。

方舱内原有的厕所和沐浴区仅供身体健康的医务工作人员和后勤保障人员使用，或临时关闭。

2.4 方舱庇护医院消防与无障碍设计

改建后方舱庇护医院内容纳的人数，应根据现有疏散楼梯及安全出口的疏散宽度确定，疏散楼梯间或高大空间安全出口净宽度按当地消防规范或按100人不小于1米计算。

主要出入口及内部医疗通道应有到达各医疗部门的无障碍通道。既有建筑内部通道有高差处，宜采用坡道连通，坡度宜符合无障碍通道要求，并确保移动病床及陪护人员同时通过的必要宽度。

2.5 方舱庇护医院配套辅助用房设计

病人入口要设置个人物品的寄存、消毒、安检用房和患者男女更衣室等。转院患者和康复患者的出口，要有消毒和打包区域。此外还可在病区附近设置紧急抢救治疗室、处置室、备餐间、被服库、开水间、污洗间、生活垃圾暂存间（污洗间、暂存间宜靠外墙，并邻近污染物出口）等用房。可在医护清洁工作区设置配液（药）室、药品库房、无菌物品库、备餐间、休息值班室、办公室等。

2.6 方舱庇护医院设计案例参考

在人口较密集的大中型城市，可以征用体育馆、展览馆、车站候车大厅、空置厂房和学校综合场馆等大型场馆建设方舱庇护医院，下面一一列举。

2.6.1 “单层展览馆式”方舱庇护医院

案例：单层展览馆式的方舱庇护医院，我们以位于武汉市东西湖区的武汉客厅中国文化博览中心改建而成的卓尔（武汉客厅）方舱医院为例（图2-1）。卓尔（武汉客厅）方舱医院由A、B、C连在一起的三个高大空旷的展览馆为基础搭建而成，属于单层（平面图见图2-2）。

首先对选用的既有建筑物——武汉客厅中国文化博览中心进行详细查阅图纸和现场踏勘，快速制订可行性方案和测算各项技术指标，各专业人员共同会审、讨论并确认方案。

图2-1　卓尔（武汉客厅）方舱医院鸟瞰图

卓尔（武汉客厅）方舱医院根据功能合理划分为三区两通道（详细参见第二章《方舱庇护医院工程设计》2.1　方舱庇护医院功能分区），利用武汉客厅中国文化博览中心平面进行功能改造：为满足方舱庇护医院收治患者的功能需求，对主展厅（A展厅）和两个侧展厅（B、C展厅）进行功能改造。其中包括：污染区，主要是护理人员工作区和收治病人区；清洁区，为医护人员生活区和物资保障区；中间的过厅，为卫生通过区。换班后的医护人员的生活住宿安排在周边其他地方，在满足隔离两周后无状况方可离开（功能分区见图2-3）。

改造要点一：利用展厅的高大空间，设置了收治病床区和护理工作区，两者功能相对独立且流线互不交叉，呈鱼骨形布局。护理工作区居中布置，向两侧可直通不同的收治病床区，患者在外侧按照入院处置→外围走廊（患者服务通道）→病情痊愈→出院清洁区→出院的流程进行治疗（功能分区见图2-4）。

改造要点二：设置医护生活区和物资保障区。利用原有的货运出入口，将周边区域设置为库房储藏区域，作为紧急情况下的物资存放；另外一侧的独立出入口作为医护出入口，设置有临时值班、办公、会议、远程会诊等功能区，并通过卫生通过区与护理工作区相连（功能分区见图2-5、图2-6）。

2.6.2 “多层展览馆式”方舱庇护医院

案例：多层展览馆式方舱庇护医院，我们以位于武汉市江汉区的武汉国际会展中心改建成的卓尔（江汉武展）方舱医院为例。武汉国际会展中心是多层展览馆，与卓尔（武汉客厅）方舱医院一样，这里也是以高大空旷的展览大厅为基础搭建，不同的是，卓尔（江汉武展）方舱医院是分上下两层，一楼是清洁区或半污染区，二楼是污染区（平面设计图纸见图2-7、图2-8）。

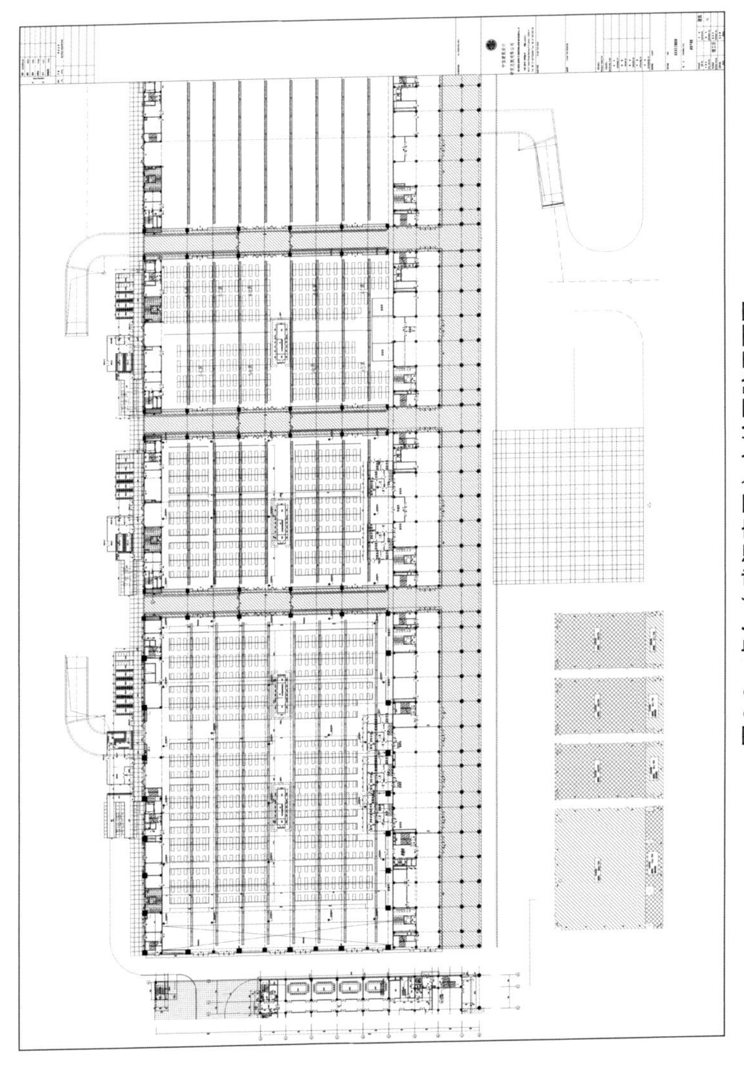

图2-2　卓尔（武汉客厅）方舱医院平面图

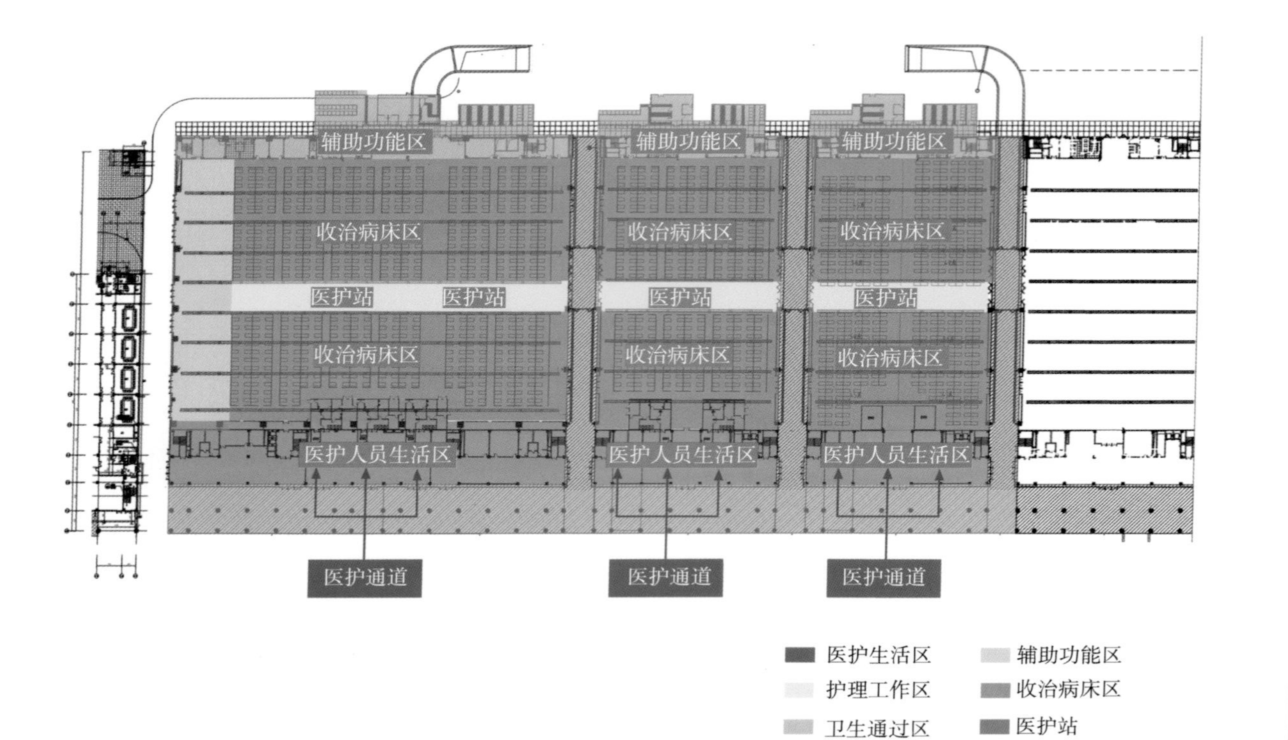

图 2-3 卓尔（武汉客厅）方舱医院功能分区图

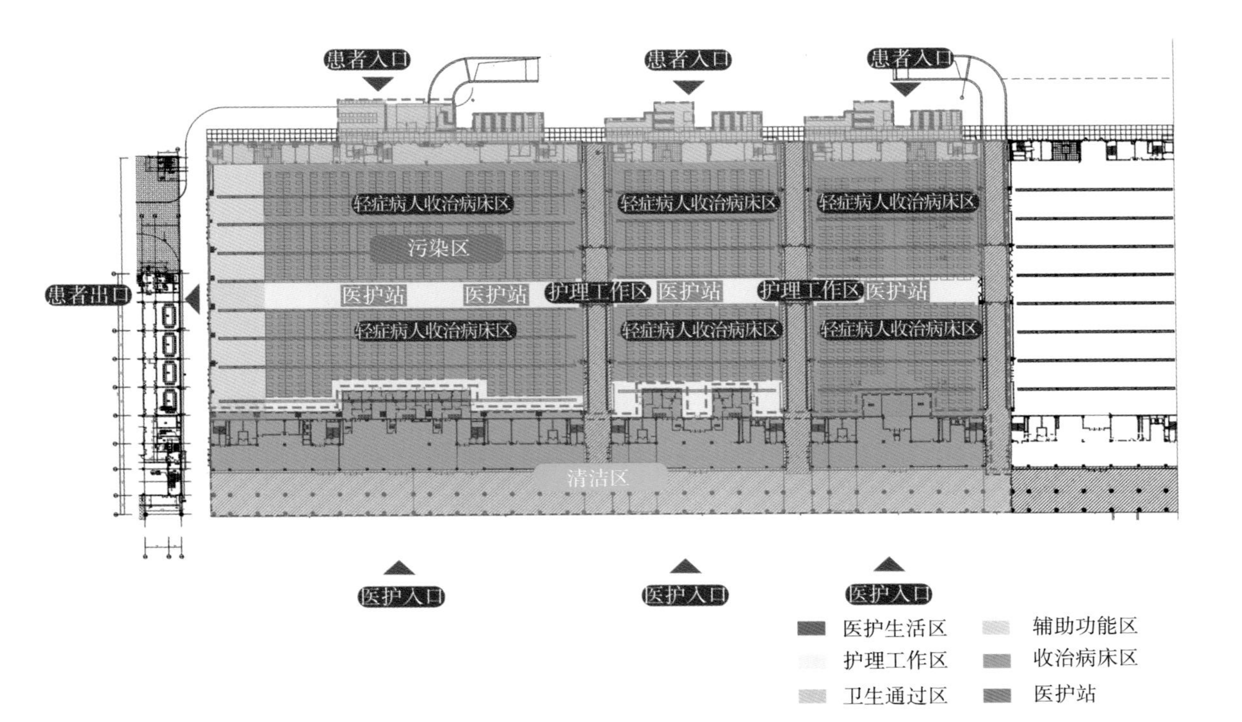

图2-4　卓尔（武汉客厅）方舱医院功能分区图

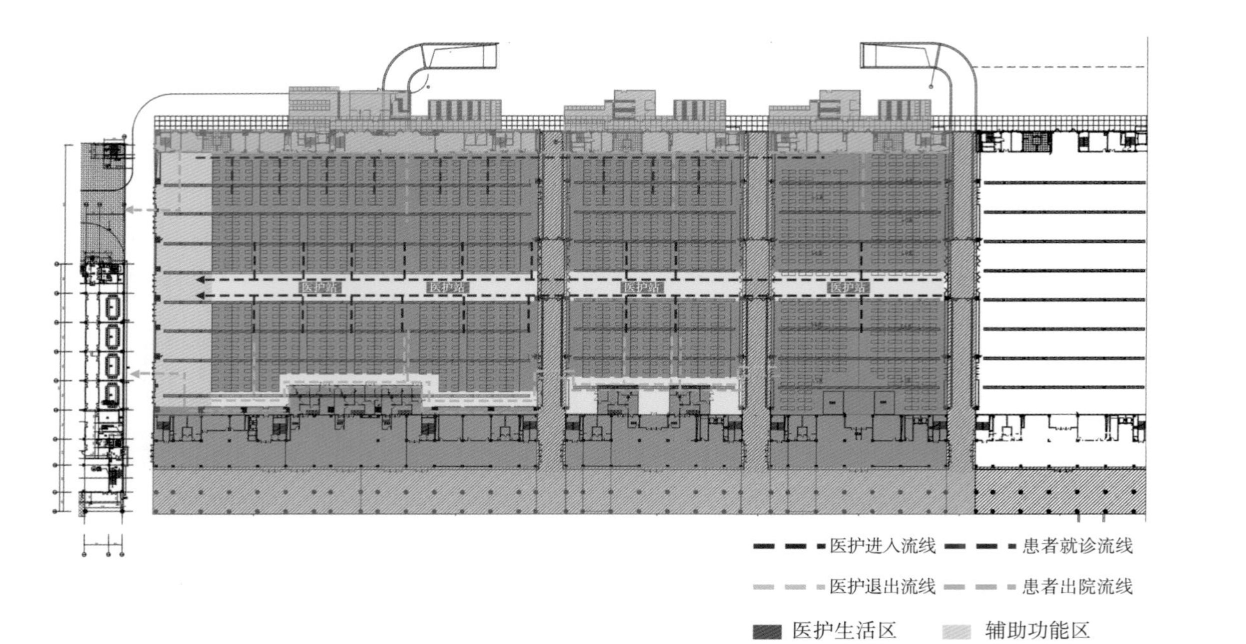

图2-5　卓尔（武汉客厅）方舱医院功能分区图

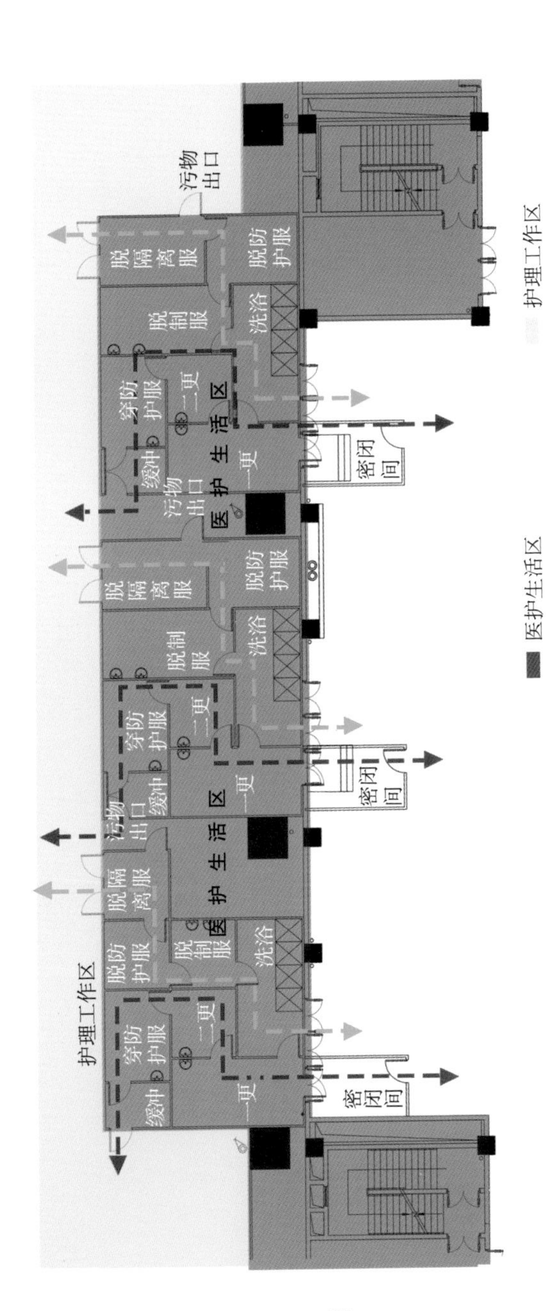

图2-6　卓尔（武汉客厅）方舱医院功能分区图

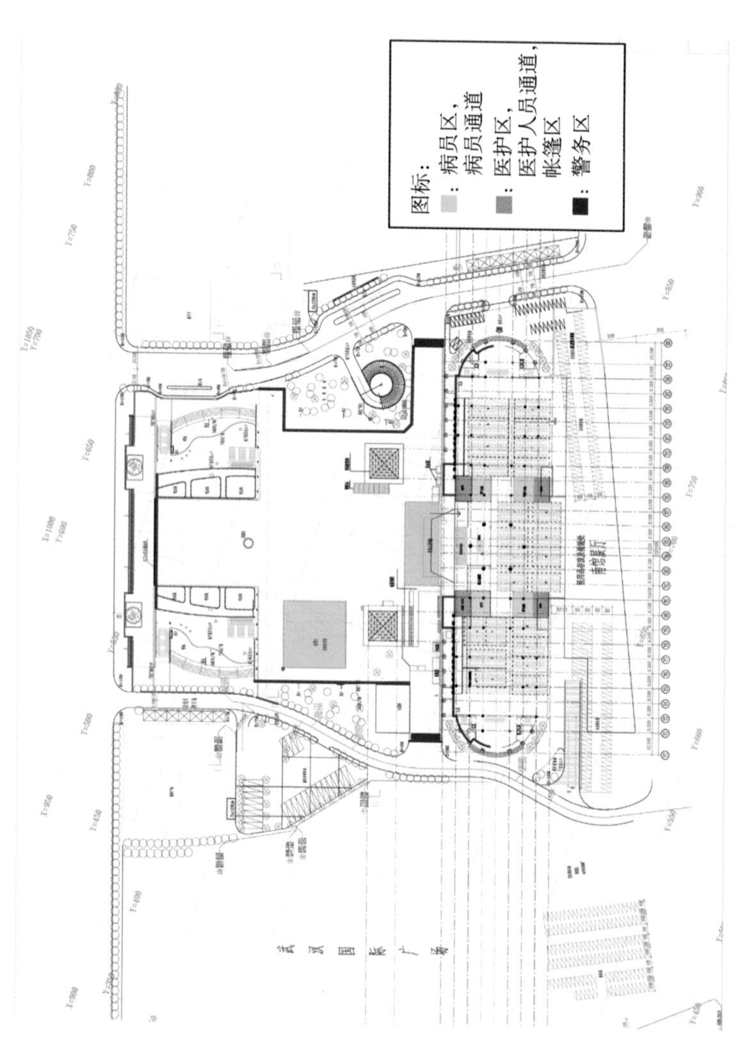

图2-7　卓尔（江汉武展）方舱医院一楼平面图

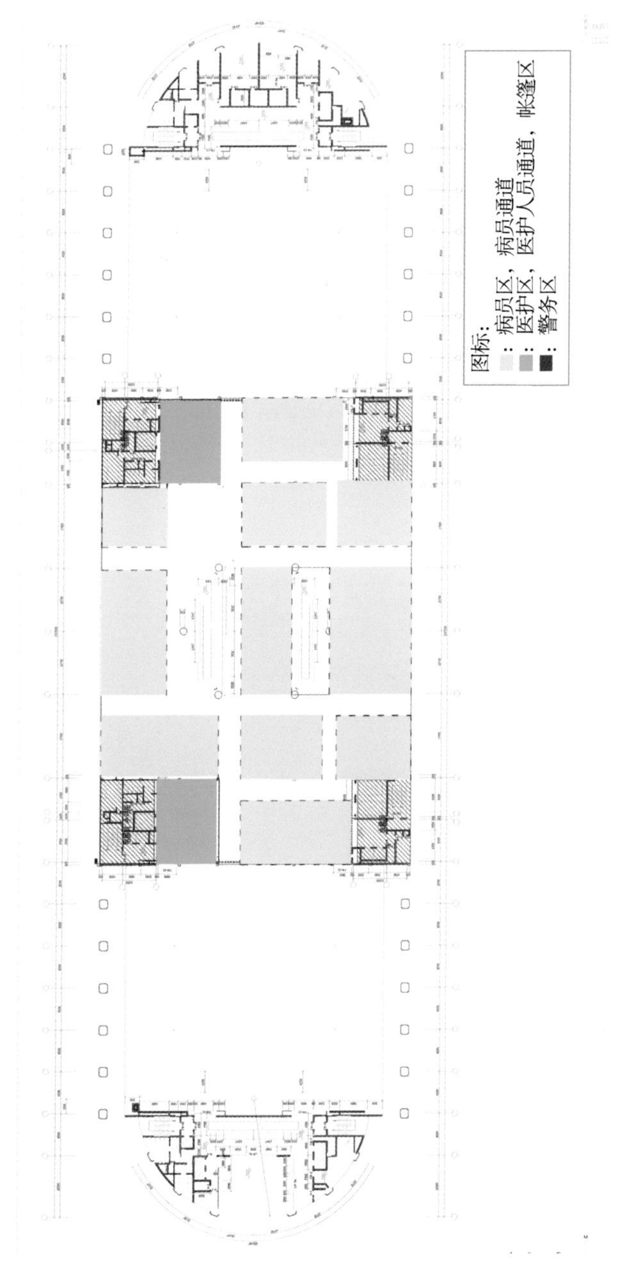

图2-8　卓尔（江汉武展）方舱医院二楼平面图

2.6.3 “体育馆式”方舱庇护医院

案例：体育馆式方舱庇护医院，我们以位于武汉市武昌区的洪山体育馆改建成的武昌方舱医院为例，它是利用一个大型室内篮球场为基础改建而成（设计图纸见图2-9 ～图2-14）。

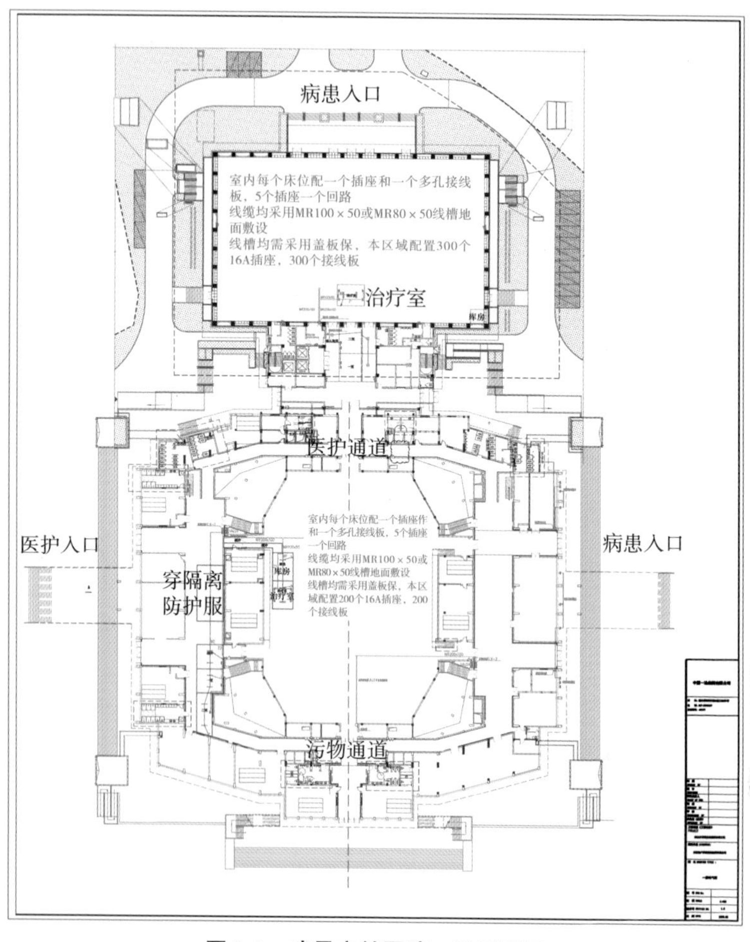

图2-9　武昌方舱医院一层平面图

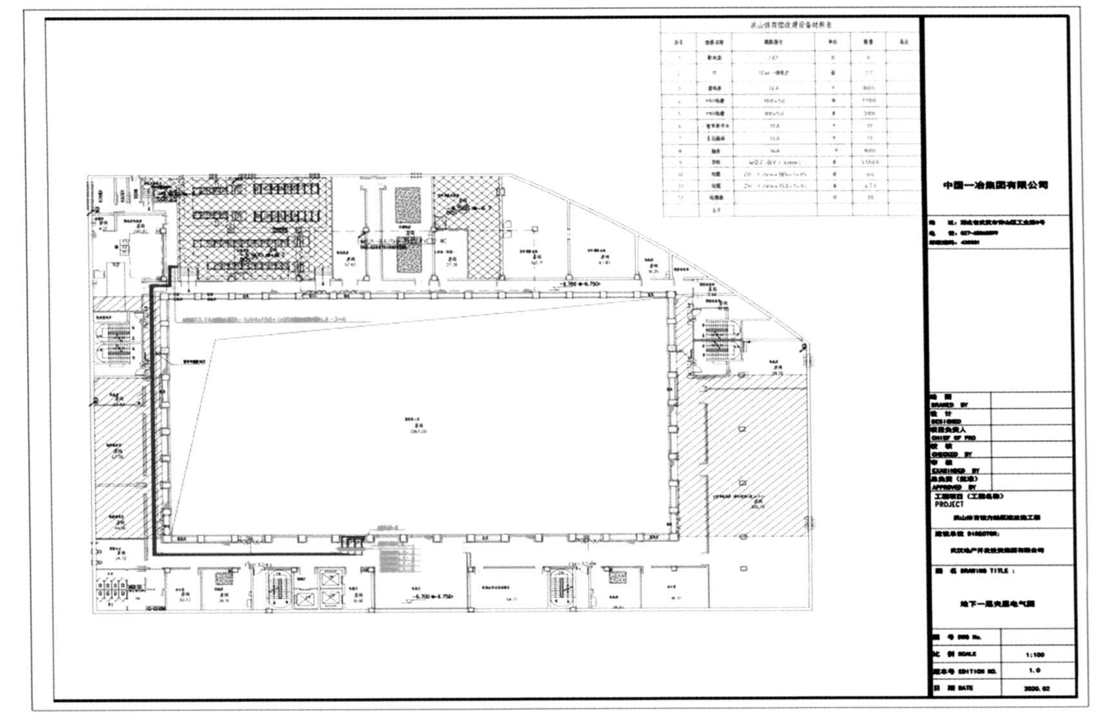

图2-10　武昌方舱医院地下一层夹层平面图

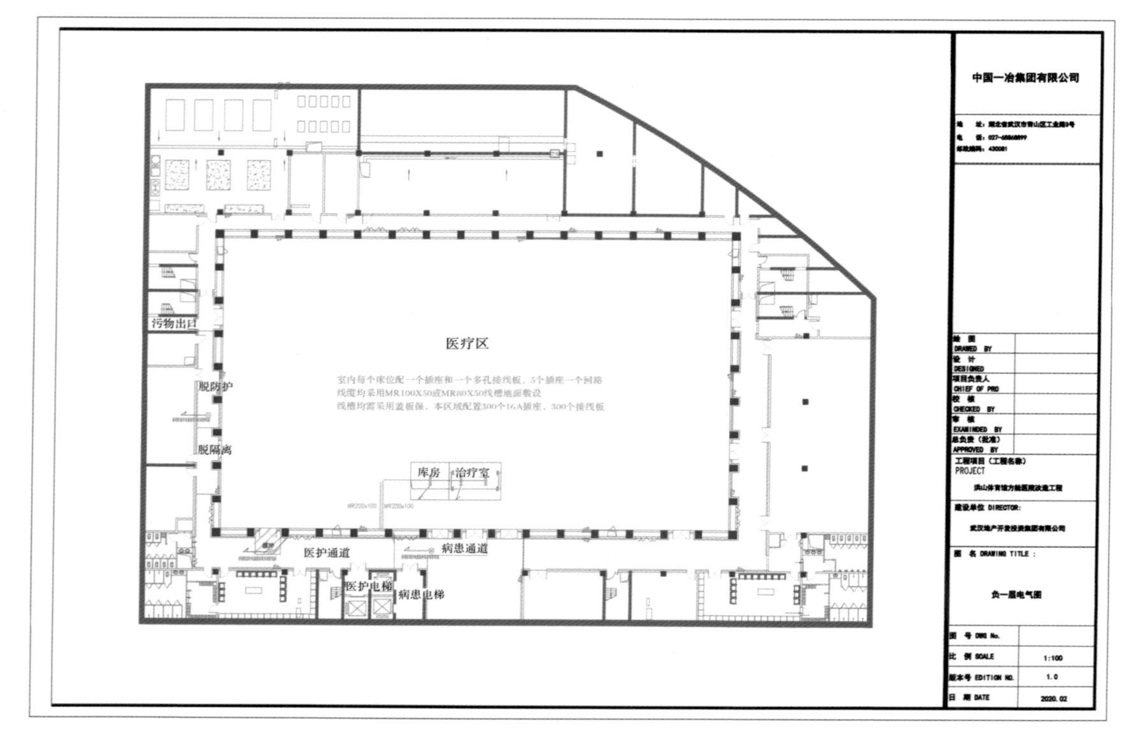

图2-11 武昌方舱医院负一层平面图

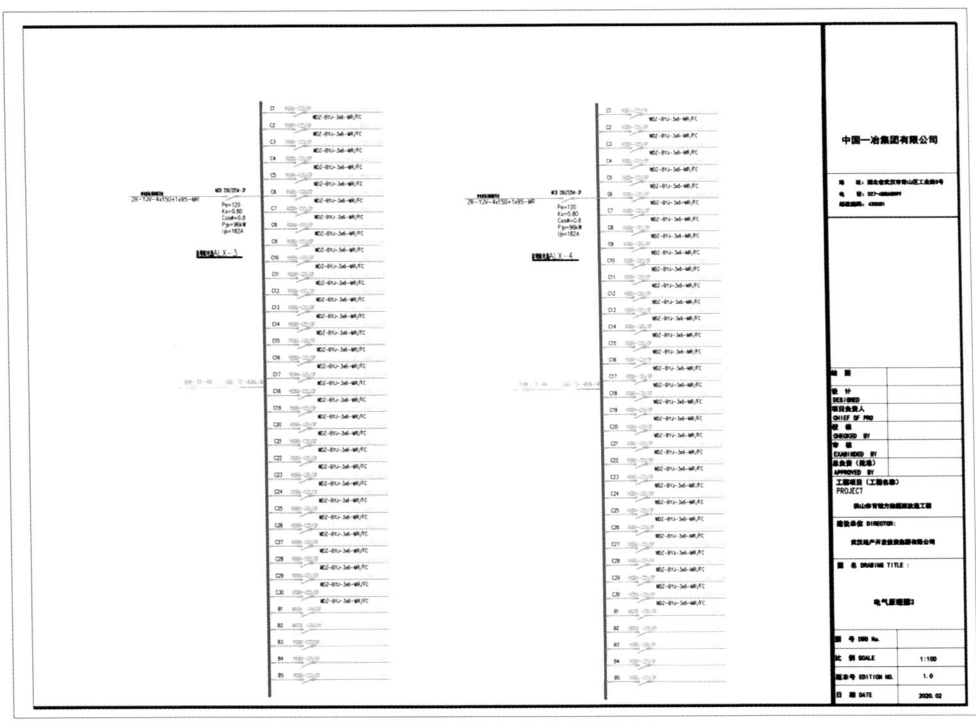

图2-12　武昌方舱医院电气原理平面图

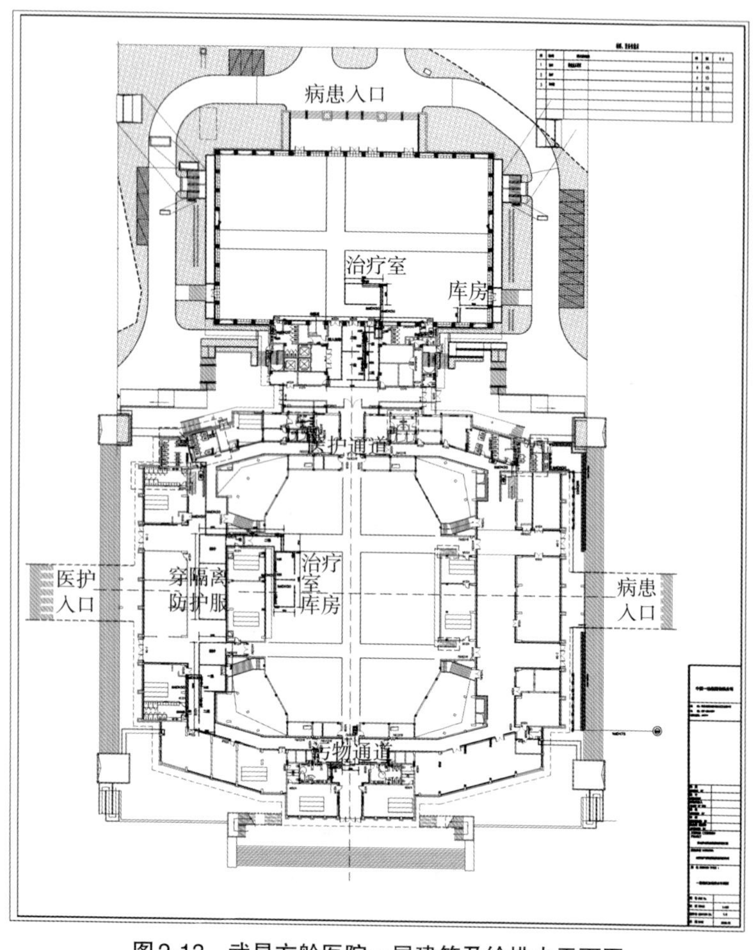

图2-13　武昌方舱医院一层建筑及给排水平面图

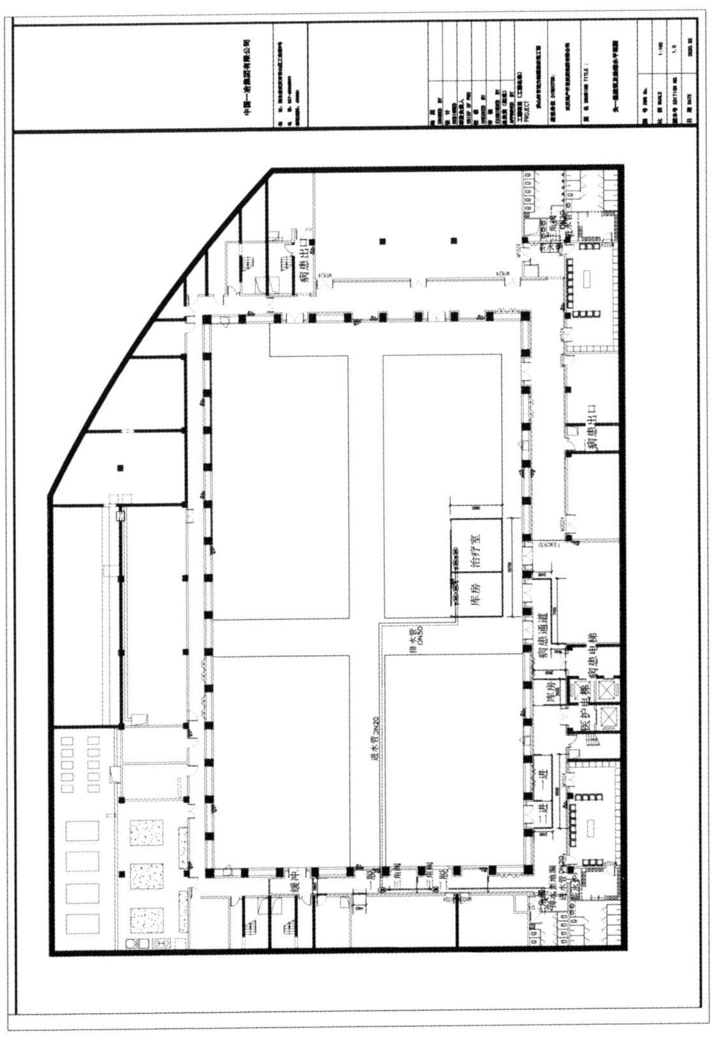

图2-14　武昌方舱医院负一层建筑及给排水平面图

2.6.4 “空置厂房式”方舱庇护医院

案例：空置厂房式方舱庇护医院，我们以武汉市长江新城卓尔康复驿站为例。长江新城卓尔康复驿站是由10个闲置厂房改建而成的10个“小方舱”组成的，每个“小方舱”占地面积1200平方米（平面设计图纸见图2-15）。

2.6.5 “候车大厅式”方舱庇护医院

案例：汽车站、火车站候车大厅改建为方舱庇护医院，我们以卓尔（汉口北）方舱医院为例。卓尔（汉口北）方舱医院是在武汉市黄陂区汉口北客运总站候车大厅的基础上改建而成（平面设计图见图2-16）。

2.6.6 “学校综合场馆式”方舱庇护医院

案例：学校综合场馆式方舱庇护医院是以学校内的综合场馆为基础改建而成，我们以位于武汉市汉阳区的汉阳方舱医院为例。该场馆是在武汉市体育运动学校综合馆（三层，1.3万平方米）和网球馆（0.48万平方米）的基础上改建而成（设计图见图2-17～图2-20）。

图 2-15　长江新城卓尔康复驿站平面设计图

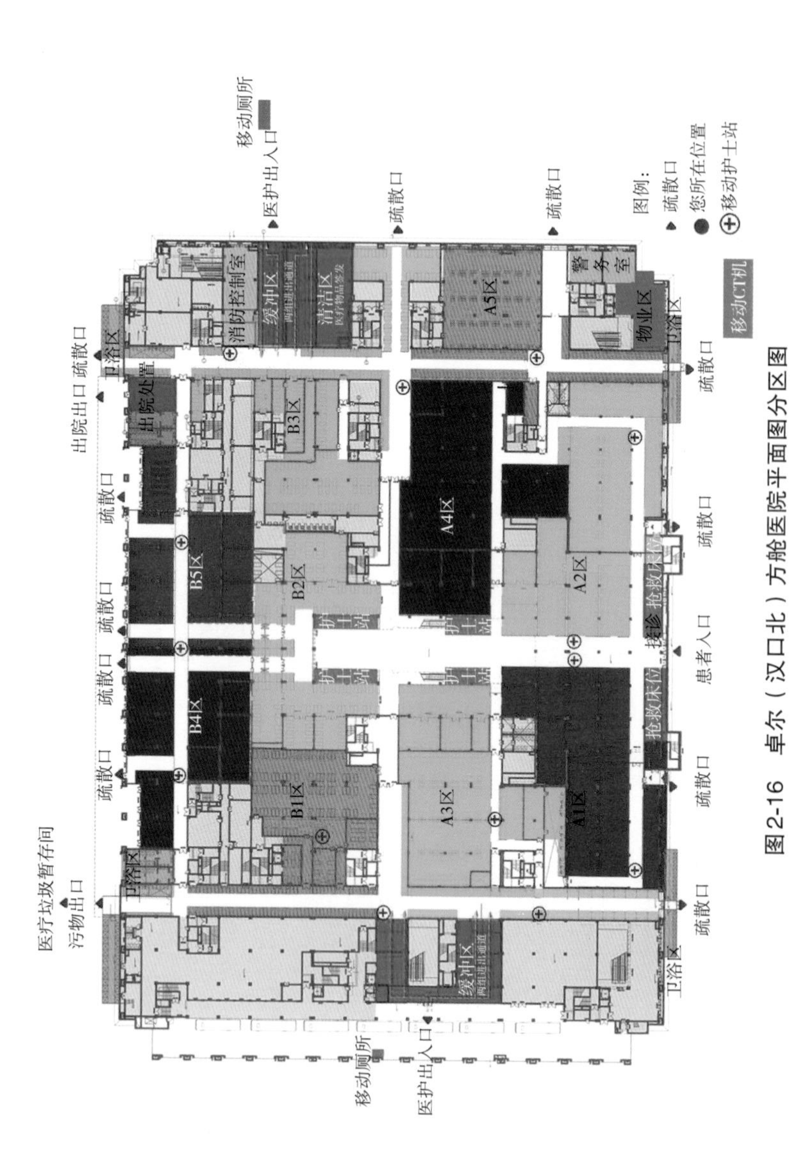

图2-16 卓尔（汉口北）方舱医院平面图分区图

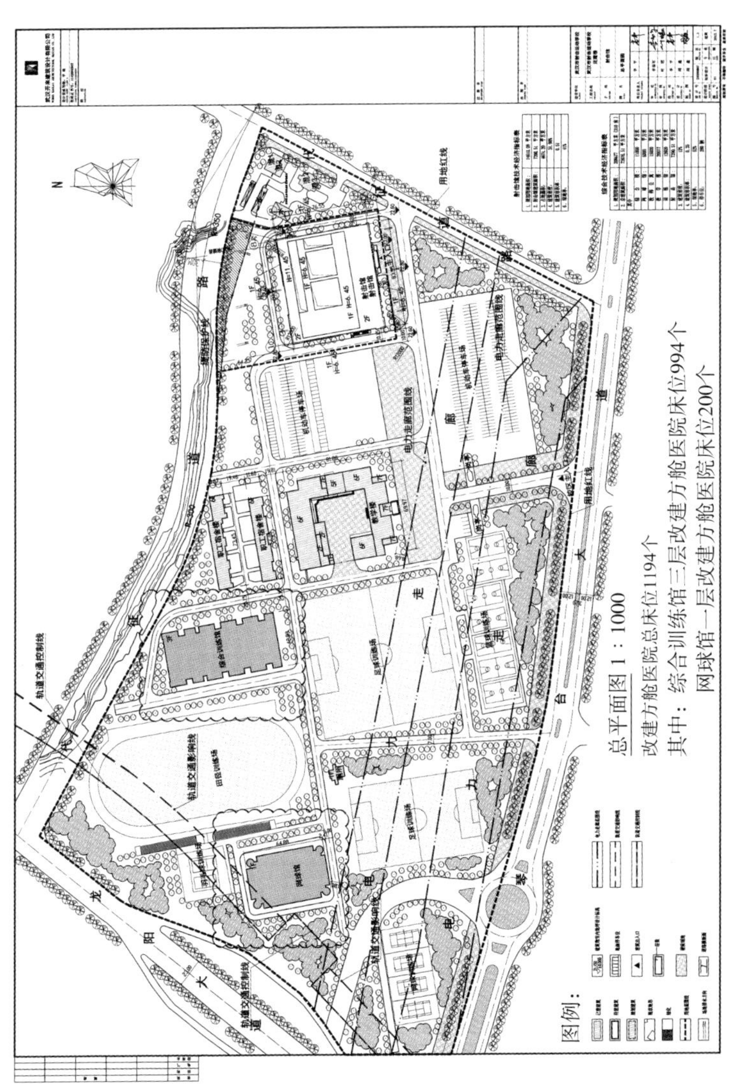

图2-17　汉阳方舱医院平面图

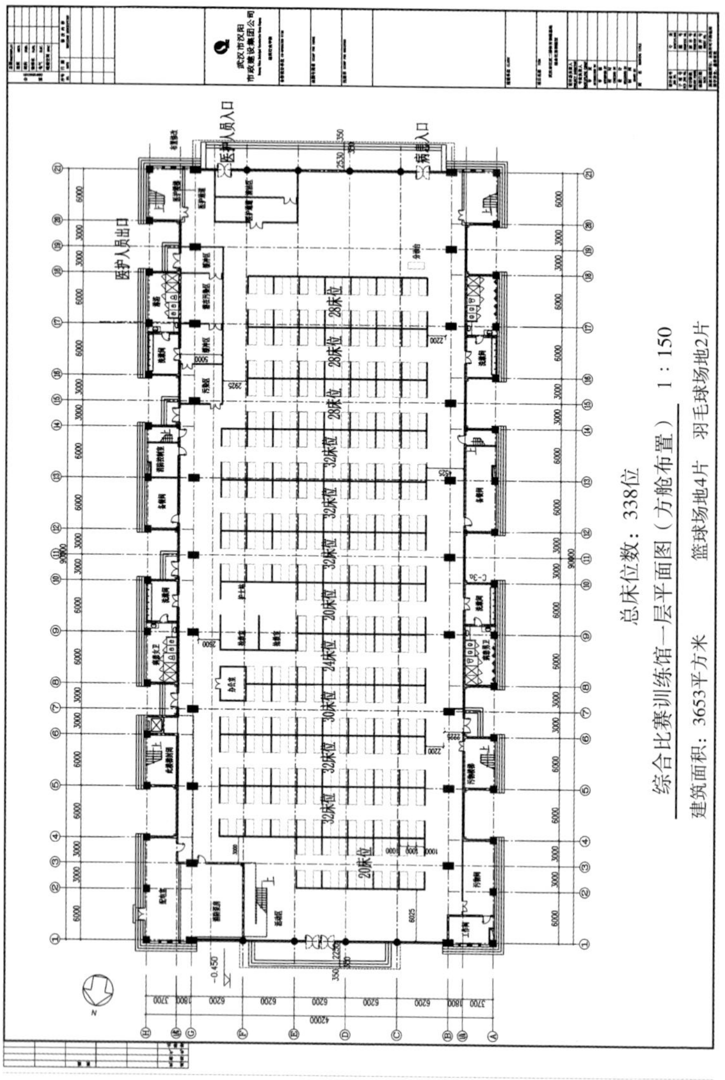

图2-18 汉阳方舱医院一层平面图

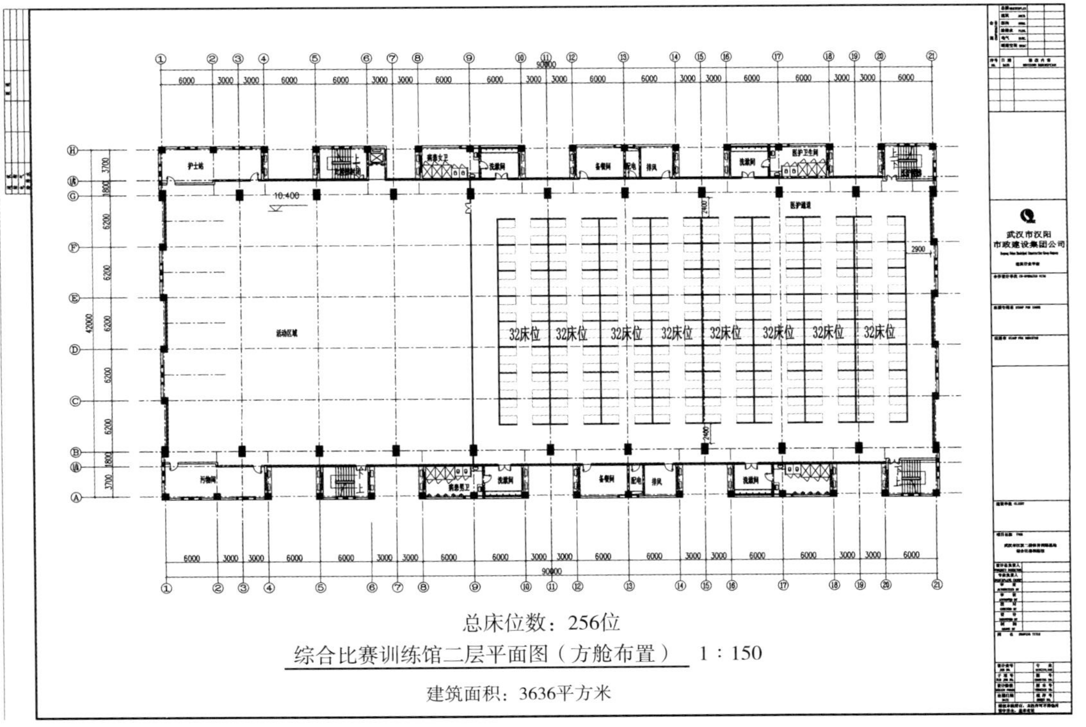

图2-19　汉阳方舱医院二层平面图

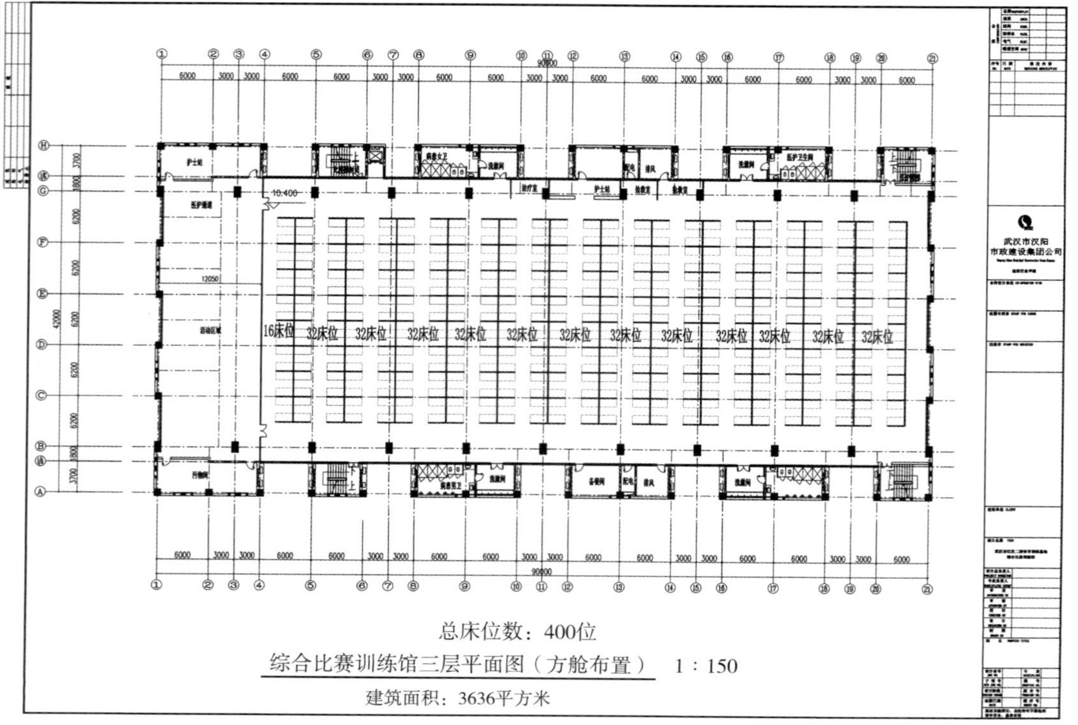

图2-20 汉阳方舱医院三层平面图

第三章　方舱庇护医院工程改造

3.1　方舱庇护医院改造内容

3.1.1　基本内容

方舱庇护医院的改造内容包括室外市政设施、污水处理设施、建筑内部分隔、建筑内部设施设备、对外交通通道、人员物资进出运输通道、相邻环境防护与改善、卫生防疫、生物安全、安全防护等方面。

改造后至征用结束期间，该建筑只能作为方舱庇护医院使用，不得兼作他用。

改建后的方舱庇护医院应满足各级卫生部门、疾控部门的要求。

既有建筑如不满足“被改造建筑的要求”，应适当改建以适应需要。

3.1.2　案例参考

展览馆、体育馆、候车大厅、学校综合场馆和空置厂房等高大空间要在短时间内变成方舱庇护医院，需要在给排水、暖通、电气和床位分区等方面进行改造。下面，我们以卓尔（武汉客厅）方舱医院为例，具体说明。

（1）A馆、B馆、C馆室内约1500张床位强电布线，每张床位配置插座；室内隔断安装。

（2）A馆、B馆、C馆室内护士站（医废间、库房、治疗室）土建及电气安装。

（3）A馆、C馆室内半污染区（一更、二更、缓冲、穿防护服、脱防护服等）机械送排风系统，并为送排风机配电，制作配电箱和敷设电气线缆等，各半污染区入口处增加密闭间（A馆、C馆共5处）。

（4）A馆、B馆、C馆室外布置盥洗室、淋浴间。其中：A馆室外盥洗集装箱4个（内含立柱洗面盆40套，小厨宝热水器40套），淋浴间集装箱2个（内含淋浴器12套，电热水器12套）；B馆、C馆室外盥洗集装箱4个（内含立柱洗面盆40套，小厨宝热水器40套），淋浴间集装箱2个（内含淋浴器12套，电热水器12套）。为以上场所配置照明灯具及敷设线缆，为消毒设施和取暖设施配电，为室外排污泵制作配电箱和敷设线缆。

（5）方舱室外给水系统（*DN*100的PE主给水管，接各用水集装箱的支管，配套阀门等）改造。

（6）方舱室外排水系统［*DN*150的UPVC主排水管，A、C区地下车库入口处75立方玻璃钢化粪池各一个，A、C区污水提升设备各一套（共4台潜污泵、2套控制柜、相应的阀门配件）］改造。

（7）A馆、C馆室内机械排风系统、电气排风机制作配电箱和敷设电气线缆。

（8）A馆、B馆、C馆病床区用生态装饰板隔断，满足防火、美观及保护隐私的要求，并安装插座，敷设线缆。

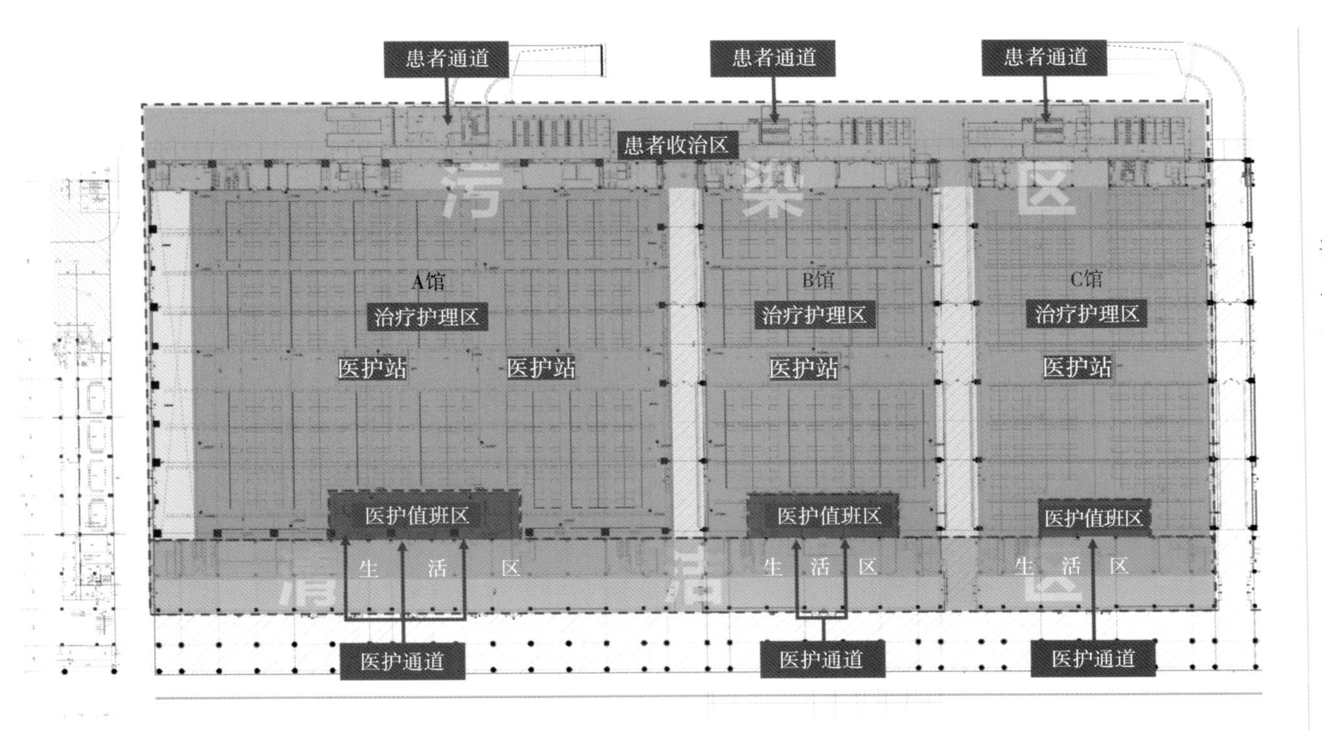

图3-1　卓尔（武汉客厅）方舱医院平面布置图

3.2 方舱庇护医院改造要求

3.2.1 选址原则要求

用于改造为方舱庇护医院的建筑宜为单层或多层建筑，建筑结构、耐火等级、防火分区、安全疏散、消防设施和消防车道等满足现行规范的相关要求。

方舱庇护医院选址应尽量远离居民区、商业区、学校等城市人群密集的活动区，远离易燃易爆有毒有害气体生产储存场所。应在医院外围设置危险标识，既有建筑与周边建筑物之间应有不小于20米的绿化隔离间距。当不具备绿化条件时，其隔离间距应不小于30米。

被改建建筑入口处应有停车以及倒车场地，能满足救护车辆的快速抵达以及快速撤离，做到对外交通便捷、内部联系顺畅、基本医疗保障设施齐全、无障碍设施齐全，并为临时停车和物资周转留出场地，用地周边宜有较为完备的安防设施。场地应有搭建临时房屋或帐篷，停放移动检验室、移动CT室等临时医疗设施，以及临时厕所、盥洗和相应的污水处理设施的空间。建筑内部空间便于迅速搭建隔断，可选择如会展中心、体育馆、大型厂房、仓库、学校综合场馆等设施设备以及消防基础条件较好的建筑。

3.2.2 结构安全要求

在方舱庇护医院的改造和建设过程中，凡涉及使用荷载可能超过原楼面设计的荷载时，结构设计人员应取得相关荷载资料据实进行复核，并根据复核结果采取相应措施。重点注意如下方面：

（1）有较重的医疗设备时，应根据设备荷载信息及其平面布置图进行复核，并根据复核结果分别采取不处理（设备荷载小于设计活荷载）、加固或更换布置位置（设备荷载大于设计活荷载）。

（2）在楼面上布置隔断时，应根据隔断的平面布置图和隔断材料的荷载信息进行复核，并根据复核结果分别采取不处理（设备荷载小于设计活荷载）、加固或采用更轻质的隔断材料（设备荷载大于设计活荷载）。

（3）当有较重的移动设备时，应根据移动设备的重量和移动路线图进行复核，并根据复核结果，采取相应措施。

（4）改建新增隔断应安装稳固，连接紧密。

3.2.3　消防设施要求

（1）原有消防设施设备能正常使用。确保应急疏散照明能正常使用。地面分区疏散指示标志设置清晰。原有安全出口满足要求，且保持畅通。

（2）室内均按严重危险级A类火灾配置手提式磷酸铵盐干粉灭火器，严重危险级场所单具灭火器最小配置灭火级别为3A，单位灭火级别最大保护面积为50m^2，选用磷酸铵盐MF/ABC5，建筑灭火器配置应按有关国家标准的规定执行。

（3）贵重设备用房、病案室和信息中心（网络）机房应设置气体灭火装置。

（4）方舱庇护医院内若增设生活给水系统，且原建筑室内消防系统未配置消防软管卷盘时，可增设消防软管卷盘或轻便消防水龙头，其布置应满足同一平面至少有1股水柱能达到任何部位的要求。

（5）为每名医护人员配备一具过滤式消防自救呼吸器，自救呼吸器应放置在方舱庇护医院内醒目且便于取用的位置。

（6）护士站宜配置微型消防站，移动式高压细水雾贮水量宜为100L。

（7）在条件许可的情况下，应确保改造后的火灾自动报警及消防联动控制系统可靠运行。

3.2.4 现场施工要求

（1）采取设计、采购、施工、验收一体化建设模式，设计、采购、施工高度融合，设计、施工等单位在施工现场密切配合，同步进行。

（2）分区、分段、分作业班组按照模块化、标准化、装配式的要求进行施工，避免班组之间交叉作业，工序之间留出合理时间间隔。

（3）按建筑平面及分区隔离布局的设计要求组织隔断墙施工，隔断墙体采用轻质防火材料，其燃烧性能不低于B1级。以分隔单元为检验批，对隔断墙刚度、强度和稳定性及连接处密封性进行检查验收。

（4）应对穿隔断墙管道和附于隔断墙内的设备采取局部加强措施，轻质隔断墙与顶棚或与其他墙体的交接处应采取防开裂措施。

（5）对通风空调、建筑电气等相关指标进行检测，确保满足设计和相关标准规范要求。

（6）加强现场作业人员的防疫安全管理。在各出入口设置固定的测温点，并且设置流动测温人员，每四小时随机进行流动测温，工作人员均应正确佩戴口罩，避免人员交叉感染。

（7）对施工场地采取通风措施，保持空气通畅。对施工期间的卫生间和办公场所每6小时进行一次消毒。

（8）施工现场严禁吸烟。加强施工场地的消防安全管理，减少明火作业，并按消防要求设置灭火器或微型消防站。

（9）设置双回路备用电源，分区设置漏电保护器，做好施工用电和运营阶段的用电安全。

3.3　方舱庇护医院给排水改造

3.3.1　设计依据

建设单位提供的条件图和有关设计资料；选用的国家现行的有关规定和地方法规。

3.3.2　给水系统

水源采用方舱现有自来水管网直供，接入管入口增设减压型倒流防止器，防止回流污染，或采用断流水箱供水。供水水压不低于水压0.25MPa，预留生活加压和生活加氯接口；在车辆停放处，应设冲洗和消毒设施；室内给水管采用PPR管，管系列等级S3.2，热熔连接；室内热水管采用PPR管，管系列等级S2.5，热熔连接；给水管上$DN<50$mm时，采用铜芯截止阀，$DN \geqslant 50$mm时采用铜芯闸阀；热水管上应采用耐高温型阀门；压力排水管上的阀门采用铜芯球墨铸铁外壳闸阀，公称压力100MPa。

3.3.3　热水系统

淋浴间采用电热水器制备热水，应选用具有接地保护，防干烧、防超压、防高温装置，有漏电保护和无水自动断电功能的产品。

3.3.4 开水系统

每个病区应单独设置饮用水供水点，供水点应足额提供常温直饮水、开水。生活用水水质应符合当地《生活饮用水卫生标准》；开水系统也可采用瓶装水饮水机。

3.3.5 排水系统

病区对外弃置的粪便、呕吐物和污、废水必须进行杀菌消毒，不得将固体传染性废物、各种化学废液弃置和倾倒、排入下水道。严禁未经消毒处理或处理未达标的病区污水、医疗污水、病区污物排放。

临时移动厕所的生活污水与洗浴区生活排水，必须经过消毒处理。污水必须集中消毒处理，参考当地医院污水处理相关要求处理，处理后的水质应符合当地现行的“医疗机构水污染物排放标准”。具体做法：

（1）移动公厕使用后，第一时间投放消毒片剂（过氧乙酸、次氯酸钠、漂白粉），水由当地环保部门负责收集后运转至污水处理站消毒处理，不得直接排入院区排水管网；

（2）病区废水就近收集后，以最短的距离就近排入院区现状化粪池，医护区污废水可排入院区现有排水检查井，病区废水与医护区污废水不得共用现状排水管网，需分别排至现状化粪池；

（3）方舱庇护医院产生的污水采取二次消毒法。对于设置有三格式化粪池的，应在第一格进行第一次投药消毒杀菌处理，并至少接触15小时禁止污水直接排放或处理未达标排放，并在接市政管网入口处（污水井）进行二次消毒杀菌处理，达标后方可排至市政污水管网，处理排放标准由环卫部门提出具体要求；

（4）投药消毒处理采用液氯、二氧化氯、次氯酸钠、漂白粉或漂白精等消毒剂实施消毒时，消毒接触池的接触时间不应低于15小时，余氯量大于6.5mg/L（以游离氯计），粪大肠菌群数少于100个/L，参考有效氯投加量为50mg/L，若难以达到前述接触时间要求，投氯量与余氯量还应适当加大；

（5）对于未建设任何污水处理设施的，应因地制宜设置临时性污水处理罐（箱）或移动式化粪池，按照有关部门的要求有效处理污水；

（6）移动厕所通气管在屋面顶部设高效过滤器或紫外线消毒，由生产移动厕所的厂家配套供应；

（7）车辆冲洗和消毒废水应排入污水系统，排水口下应采取水封措施，水封深度不得小于50mm，严禁采用活动机械活塞替代水封。

3.3.6　排水管道安装

排水管应采用UPVC排水管，塑胶粘结。卫生间采用带滤网的直通式地漏或网框地漏，下设存水弯，其水封高度不小于50mm。构造内无存水弯的卫生器具与生活排水管道或其他可能产生有害气体的排水管道连接时，在排水口以下应设存水弯，其水封深度不得小于50mm。排水管道除图中注明外，按下列坡度安装。

排水管的横管与横管、横管与立管必须采用45度三通，45度四通或90度斜三通，90度斜四通。

方舱庇护医院排水管道坡度安装标准

管径	*DN*75	*DN*100	*DN*150	*DN*200
污水、废水管标准坡度	0.025	0.020	0.02	0.01

3.3.7 卫生器具

所有用水点及其他有无菌要求或需要防止院内感染场所的卫生器具处均应采用非接触性或非手动开关，并应防止污水外溅。采用非手动开关的用水点应符合下列要求：

（1）医护人员使用的洗手盆，以及细菌检验科设置的洗涤池、化验盆等，应采用感应水龙头或膝动开关水龙头；

（2）公共卫生间的洗手盆应采用感应自动水龙头，小便斗应采用自动冲洗阀，坐便器应采用感应冲洗阀，蹲式大便器宜采用脚踏式自闭冲洗阀或感应冲洗阀；

（3）洗手盆采用感应水龙头或者脚踩出水，蹲便器采用脚踏阀。

3.4 方舱庇护医院给通风空调改造

3.4.1 通风工程实施的必要性

方舱庇护医院的既有建筑通风与空调系统大多为正压系统，少数为负压系统，无法满足医疗要求。通风系统改造就是最大程度利用原有通风空调设备设施，根据使用需求和设计要求增设部分设备设施，改变原系统送风排风运行策略，使清洁区、半污染区和污染区之间形成压力梯度，保证清洁区压力最高，半污染区次之，污染区压力最低。

3.4.2　通风工程实施模式

为加快工程进度，同时充分实现设计效果，可采用“设计—采购—施工”一体化的工程总承包模式（EPC）。设计、采购、施工人员同空间、同时间、平行化的沟通讨论，做到设计方案、设备货物、施工方案同时确定，设计出图、设备调配、施工人材机多环节同时开展，以空间换时间压缩工期节约成本。设计阶段，采购团队、施工团队与设计师团队精准对接，及时沟通，提前踏勘现场、订购设备、准备施工材料、组织施工班组、制定施工方案，设计初步完成即具备施工条件；施工阶段，设计师驻场服务，根据现场条件及时调整设计方案。

3.4.3　通风工程设计要点

（1）污染区和半污染区应以机械通风方式为主，排风机入口处均加装粗中高效（或亚高效）过滤器。清洁区等小空间可采取机械通风方式或自然通风。

（2）污染区和半污染区集中空调系统应使用空气净化消毒装置。有条件时空调机组可设置亚高效过滤器以上等级的洁净空调系统；可在回风过滤器、表冷器附近安装紫外线消毒灯。

（3）应根据设定的医护人员区域和隔离病房区域做好临时进、排风设置，气流流向为从医护人员区域至病房区域，送、排风机（口）的设置位置应形成合理的气流通道，尽量保证不留通风死角。

（4）原有空调和排风系统可以利用时，应设置为直流式送、排风系统，空调机组关闭回风阀，新风阀全部开启，全新风送入，排风量应大于送风量（排风机风量不够时，可开启排烟风机）。原有空调和排风系统无法利用或未设置通风系统的，应增设通风系统。需临时加装排风系统时，宜选择风量、风压合适的风机箱，设置高度不高于2m，并设置防护措施。通风系统要求24小时不间断运行。

（5）排风量应按每人不小于150m^3/h设计。

（6）医护人员通过“一次更衣→二次更衣→缓冲间”后，从清洁区进入到污染区，在“一次更衣”设置不小于30次/小时的送风，各相邻隔间设置D300通风短管，气流流向从清洁区至隔离区。医护人员通过“缓冲间→脱隔离服间→脱防护服间→脱制服间→淋浴间→一次更衣”后，从隔离区返回清洁区，在“缓冲间→脱隔离服间”设置不小于30次/小时的排风，各相邻隔间设置D300通风短管，气流流向从清洁区至污染区。

（7）每个隔离病房区域，设置若干台具有杀菌消毒功能的空气过滤器，根据实际情况需要设置升温设施的可设置若干台电热油汀。

（8）隔离病房区域应采用应急干厕，隔离病房盥洗间、医护人员区域设置的厕所应增设排风机，满足换气次数12次/小时，排风机入口宜加装高效过滤器。

（9）应根据实际情况设置送、排风机的安装位置，应确保新风取自室外，新风取风口及其周围环境必须清洁，保证新风不被污染。室外排风宜高空排放，且与任何进风口水平距离不得小于20m，垂直距离不得小于6m。

3.5 方舱庇护医院电气及智能管理改造

（1）通风设备控制箱，宜采用成套定型产品，并由护士站（值班室）集中控制。

（2）为减少原建筑顶部照明灯具的眩光影响，可在大开间周边墙上增设一些照明灯具，或地面增设一些立杆灯，增设的灯具宜带不透明罩或采用间接照明，电源由备用回路引来。

（3）应提供足够的无线网络接入条件，保证4G/5G网络全覆盖。有条件的场所，宜增设无线AP、WiFi全覆盖。

（4）地面增设的照明或插座电源线路及弱电线路应采用金属管或金属线槽敷设，管槽的敷设应避开人员通行及货物运输通道，无法避开时应采取必要措施。

（5）在卫生间缓冲区设置紫外杀菌灯或空气灭菌器插座，电源由预留回路引来。紫外杀菌灯应采用专用开关，不得与普通灯开关并列，并有专用标识。

（6）医疗设备间、淋浴间或有洗浴功能的卫生间等处，应设置辅助（局部）等电位联结。

（7）在护士站（值班室）设置一键报警按钮，信号送至保安值班室。

（8）病患休息区、护士站宜实现视频监控覆盖。

（9）病患休息区宜分区设置投影仪及幕布，接入有线电视系统。

（10）有条件的病床，插座宜布置在隔断上。所有回路送电之前需做一次绝缘检测，低压或特低电压配电线路线间和线对地间的绝缘电阻测试电压及绝缘电阻值不应小于0.5欧，保证新增线路的安全可靠。

3.6　方舱庇护医院病区设置

病区设置排风系统维持负压，排风量按不小于200m^3/h设计。排风系统设置初效过滤器（G4）＋中效过滤器（F8）＋高效过滤器（H11），排风经过滤器后由立管高空排放，并采用开启病区外门的方式，负压补风。

风机出口的排风管采用布袋风管制作，由供货方加工好后运至现场安装。排风系统的室外排风口应不低于屋檐，在高空排放。排风机入口设置防鼠钢丝网，风机之间及与门洞的缝隙应封闭严密。

病区内及护士站、主要通道设置静电空气净化机，杀灭细菌、病毒，净化病区空气。

3.6.1 缓冲区设置

缓冲区设置送排风系统，缓冲区的一更送风量按30次/h设计，一更与二更之间的隔墙设置洞口，使气流由一更流向二更。洗浴间、脱隔离服、脱防护服间设置排风系统。

缓冲区的送排风机均内设空气处理装置（亚高效）。风管尺寸为D110，采用金属波纹管和UPVC管制作。

3.6.2 其他区域设置

室外盥洗室、淋浴间设置排风系统，排风量按不小于8次/h设计，排风机均内设空气处理装置（亚高效）。风管尺寸为D110，采用金属波纹管和UPVC管制作。

3.7 方舱庇护医院硬软件搭建

3.7.1 硬件建设

舱内搭建：舱内硬件建设主要是病床，建议调用学校或部队使用的高低床、折叠床，事后可消毒处理，用环保、无污染的生态装饰板隔开；还需要采购被褥、垫子等，如果气温低，还应该准备一些电热毯、暖手宝和插线板等，条件允许的地方可配备5G通信网络，方便舱内与舱外联系。

舱外搭建：应配备移动影像车、检验车、抢救车和P3移动实验室等高科技装备，用于检验、检测病毒。

3.7.2　软件建设

方舱内应按患者人数和方舱大小配备一定比例的医护队、清洁工和送餐员等。以卓尔（武汉客厅）方舱医院为例，前后从全国各地共调集22支医疗救援队、15支护理团队和1支放射技师团队，共计1169人。其中，医护人员实行4级管理制度：医务部主任、分厅负责人、舱组长、医生；医护人员采用四班倒、轮班制，每6小时一班，分A、P、N1、N2四班，每两天轮休一次，每班由1名医生、5名护士，负责照顾100名左右患者。

方舱外主要是为患者提供送餐、治安、清洁、消杀、心理疏导、供水、供暖、供电保障等后勤保障工作。还是以卓尔（武汉客厅）方舱医院为例，后勤保障团队共计100多人，其中54名清洁人员，每天分三班不间断清扫，平均每天处理600～700千克废弃物。

第四章　方舱庇护医院运营方案

4.1　方舱庇护医院运行管理总则

4.1.1　管理原则

定向收治、集中隔离、单元式分区管理、标准化治疗、双向转诊。

4.1.2　运行目的

集中隔离治疗各社区新冠肺炎确诊病例轻症患者，控制传染源，避免在社区产生交叉感染，统一进行疾病宣教、心理疏导，给予患者及时科学的治疗观察，防止病情加重，降低病重率和病死率。

4.1.3　组织架构

方舱庇护医院在防控指挥部门的统一调度下，院长全面负责医院运行管理工作，副院长分工协作。医院下设多个工作组开展具体工作：

（1）综合信息组（设组长）：负责运行方案制定、流程确定、总体分工和协调、运行各类信息的收集、公示和上报、转运对接与协调、行政人员排班及运行问题的协调处理。

（2）医疗运行组（设组长）：含医务组和护理组，医务组负责制定医疗方案、相关核心制度和流程的制定与落实、医生信息汇总及排班；护理组负责护理工作安排，包括制定护理方案、流程、护理人员信息汇总及排班等。

（3）院感控制组（设组长）：负责院感制度的制定与落实，防护培训，运行期间院感的巡查和督导。

（4）后勤保障组（设组长）：负责物资调配、生活保障、设备设施保障、药品准备、环境卫生、医疗废物处理、排污等工作。

（5）各医院可结合实际情况，设置职能管理部门。

4.1.4　工作方法

各工作组内明确分工，完善本组工作制度和工作方案，进行分组排班，并实行总值班制度，负责沟通、协调、信息上报、督促各相关负责组及时处理方舱运行中出现的各类问题，实行分时间段排班、交接班管理和汇总上报制度。

4.1.5　工作纪律

（1）服从总体指挥、分工明确、主动作为、相互支持。

（2）通信通畅，保持联系电话24小时开机。

（3）不得迟到早退，特殊情况提前报备。

（4）遵守信息保密制度，不得随意泄露不当信息。

4.2　方舱庇护医院患者收治标准

结合实际情况，方舱庇护医院收治的新冠肺炎确诊病例，须同时满足以下入舱条件：

（1）轻型（临床症状轻微，影像学未见肺炎表现）、普通型（具有发热、呼吸道等症状，影像学可见肺炎表现）。

（2）有自主生活能力，可以自主行走。

（3）无严重慢性疾病，包括高血压、糖尿病、冠心病、恶性肿瘤、结构性肺病、肺心病以及免疫抑制人群等。

（4）无精神疾患史。

（5）静息状态下，指血氧饱和度（SpO_2）＞93%，呼吸频率＜30次/分。

（6）需要特殊说明其他情况。

4.3 方舱庇护医院入住流程

（1）每天上午10点之前，由各患者收治区的护理组组长（护士长）根据空余床位情况，上报可转入患者数量至院办主任，院办主任与分管院长对接，确定当日拟接收患者数量，并上报方舱庇护医院疫情防控指挥部。

（2）指挥部根据方舱庇护医院提供的空余床位数及拟接收患者数量，确定转至方舱庇护医院患者数量，并将患者名单及基本信息（包括病人身份信息、联系电话、病情信息、用药信息等）发送给方舱庇护医院。

（3）方舱庇护医院组织专家组根据入院标准对患者的状况进行审核，确定当日拟收治患者名单及分配病区与床位号，并为每位患者开具转入证明，上报指挥部。

（4）指挥部打印每位患者的资料（资料上写病人编号），连同转入证明一并交患者随身携带。

（5）指挥部负责统筹安排患者转运，协调救护车调度、随车人员、随车资料等，发车时发送车号及病人编号给方舱庇护医院（图4-1）。

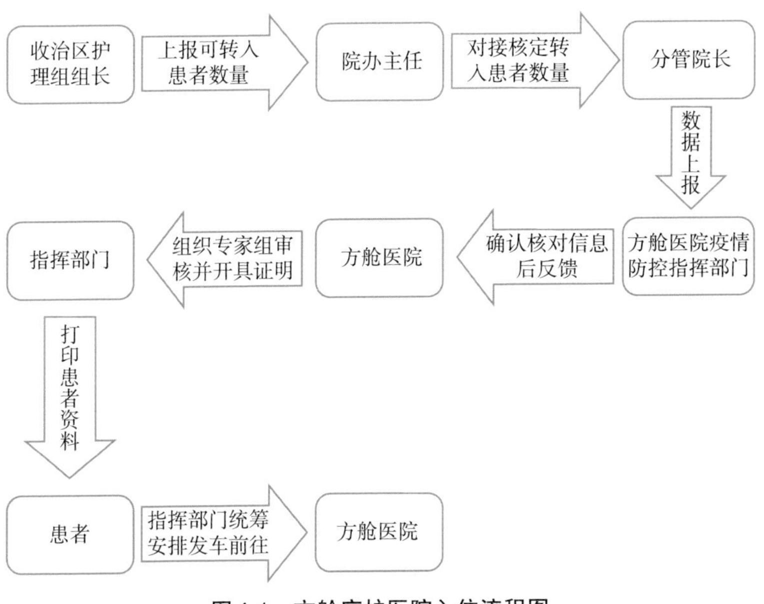

图4-1　方舱庇护医院入住流程图

4.4　方舱庇护医院预检分诊

方舱庇护医院安排医务人员对收治患者进行初步预检分诊。预检评估后，对于符合收治标准的患者，医务人员负责指引患者及时入住方舱庇护医院；对于不符合收治标准的患者，如发现病情较重，应遵循先收再转的原则。为保障医疗安全，应优先安置到舱内重症观察救治区，给予及时治疗和严密监护，并及时联系安排转定点医院救治。

4.5 方舱庇护医院住院患者日常检查

密切监测生命体征及血氧饱和度。主要是：

（1）测量体温并记录，每天4次，时间分别为8AM、12AM、4PM、8PM。

（2）记录呼吸频率，每天2次，时间分别为8AM、8PM。

（3）测量心率、手指血氧饱和度，每天2次，时间分别为8AM、8PM；当患者自我感觉不适时，可使用手指夹式血氧仪进行经皮血氧饱和度检测；较重的患者需要持续对血氧饱和度进行监测，直至病情缓解或转院。

（4）实验室检测及影像学检查由责任医生根据病人情况决定。

（5）特殊检查：由责任医生根据病人病情决定。

4.6 方舱庇护医院重症患者管理

重症患者是指入院时已是重症以及轻症患者住院期间病情加重者。要求各病区为重症患者设置相对独立观察救治区，配置氧气瓶、抢救车、抢救药品、简易呼吸器、监护抢救设备，有条件的可配置无创呼吸机、转运平车等，专人负责，加强和优先配置医护人员。

4.6.1 重症患者启动会诊、转入重症观察区的指征

（1）自觉症状持续不缓解，或有加重趋势；

（2）体温38℃内给予口服温开水、物理降温，若无缓解，体温高于38.5℃；

（3）RR≥30次/分，予以吸氧不缓解；

（4）指血氧饱和度≤93%；

（5）HR≥100次/分，BP≥140/90mmHg；有高血压者，予以日常口服降压药治疗；无高血压史，予以吸氧、退热后血压仍高者。

4.6.2　重症患者抢救流程

用轮椅或平车将患者转至重症观察救治区域；评估病情，开放静脉通道，实施救治；实施生命支持和监护；按照转运流程上报指挥部请求转至定点医院；记录情况上报。

4.6.3　重症患者转院标准

一般情况，符合以下其中一项即达到转院标准：呼吸窘迫，RR≥30次/分；静息状态下，指血氧饱和度≤93%；动脉血氧分压（PaO_2）/吸氧浓度（FiO_2）≤300mmHg（1mmHg＝0.133kPa）；肺部影像学显示24～48小时内病灶明显进展＞50%；合并严重慢性疾病，包括高血压、糖尿病、冠心病、恶性肿瘤、结构性肺病、肺心病以及免疫抑制人群等；其他特殊紧急原因需转出的。

4.6.4　重症患者转院流程

方舱庇护医院内的患者在舱内治疗期间，病情发生变化时，经方舱区域内会诊小组会诊后符合转出标准的患者，参照以下转运流程执行：

（1）责任医生经检查评估后，请方舱区域内上级医师会诊；

（2）经会诊后符合重症标准的患者立即上报指挥部，请求转至定点医院救治；

（3）填写转院登记表，等待指挥部转运指示；

（4）接到指示后协助完成患者的交接工作，并配备医护人员护送转运，并做好登记报表，上报信息。

4.7 方舱庇护医院患者出院标准和流程

4.7.1 出院标准

同时满足以下条件，方可出院：

（1）体温正常达3天以上；

（2）呼吸道症状明显好转；

（3）肺部影像学显示炎症明显吸收；

（4）连续2次呼吸道病原核酸检测阴性（采样时间间隔至少24小时）。

满足以上条件的患者，尚需经病区和医院两级专家会诊，一致认为达到出院条件后，方可出院。

4.7.2 出院流程

（1）责任医生完成出院标准相关检查后，请方舱区域内上级医师会诊；

（2）经会诊后符合出院标准的患者，上报指挥部医疗组；

（3）填写转运登记表，等待指挥部转运指示；

（4）接到指示后协助指挥部完成患者的交接及转运工作，并做好登记报表，上报信息；

（5）告知居家隔离注意事项：单人单间戴口罩隔离，尽量不要外出，居家隔离期间，每天都要测体温，隔离时间14天。如无居家隔离条件的，指挥部应安排集中隔离。当病人再次出现发热咳嗽等症状，或原有症状加重，立即报告社区相关负责人，并到附近指定医院就医。

4.8　出院病人的消毒处理流程

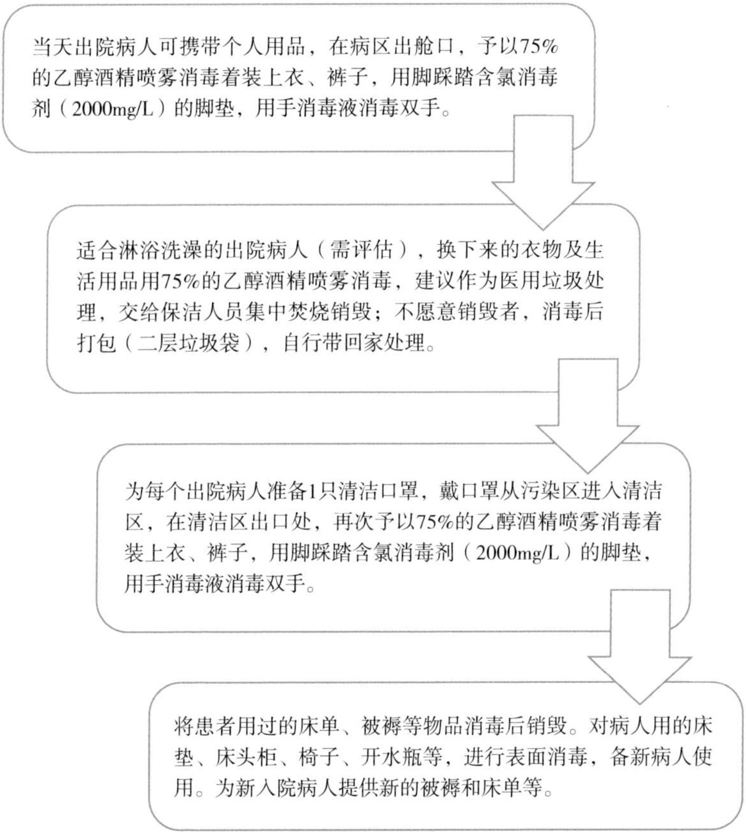

当天出院病人可携带个人用品，在病区出舱口，予以75%的乙醇酒精喷雾消毒着装上衣、裤子，用脚踩踏含氯消毒剂（2000mg/L）的脚垫，用手消毒液消毒双手。

↓

适合淋浴洗澡的出院病人（需评估），换下来的衣物及生活用品用75%的乙醇酒精喷雾消毒，建议作为医用垃圾处理，交给保洁人员集中焚烧销毁；不愿意销毁者，消毒后打包（二层垃圾袋），自行带回家处理。

↓

为每个出院病人准备1只清洁口罩，戴口罩从污染区进入清洁区，在清洁区出口处，再次予以75%的乙醇酒精喷雾消毒着装上衣、裤子，用脚踩踏含氯消毒剂（2000mg/L）的脚垫，用手消毒液消毒双手。

↓

将患者用过的床单、被褥等物品消毒后销毁。对病人用的床垫、床头柜、椅子、开水瓶等，进行表面消毒，备新病人使用。为新入院病人提供新的被褥和床单等。

图4-2　方舱庇护医院出院病人消毒处理流程图

第五章　方舱庇护医院后勤保障方案

为做好方舱庇护医院后勤保障供应服务，后勤保障从餐饮、住宿、保洁、用品、其他物资供应等方面，结合后勤实际情况，制定具体方案（图5-1）。

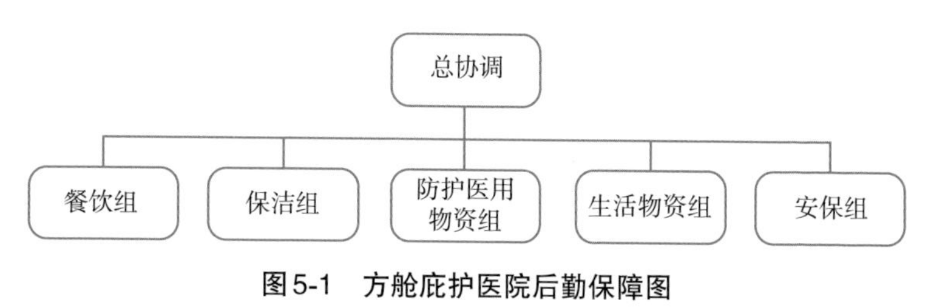

图5-1　方舱庇护医院后勤保障图

5.1　物资保障

5.1.1　餐饮保障

5.1.1.1　人员配备

负责人1人，提供联系方式，与其他各部门做好协调。

5.1.1.2　具体工作

（1）每日统计患者、工作人员人数名单，提前准备，保证适当充裕。

（2）规定时间内按照名单有序发放。

（3）与餐饮公司做好对接。

（4）做好用餐清洁、安全卫生等工作。

（5）坚持每天配送新鲜菜品，对数量、质量严格验收，保证饮食安全。

（6）根据早、中、晚三餐调配，建议早餐发放时间为7：00～8：00，中餐发放时间为11：30～12：30，晚餐发放时间为17：30～18：30。

5.1.1.3　具体要求

（1）在规定时间前，分别将患者和医务工作者的餐饮送至指定区域，并在通讯群内或通过广播其他方式通知；

（2）每日汇总一日三餐发放情况，包括早、中、晚三餐、患者领取情况、工作人员领取情况。

5.1.2　清洁保障

5.1.2.1　人员配备

负责人2人，提供联系方式，与其他各部门做好协调。

5.1.2.2　具体工作

（1）每个区域都必须安排保洁人员，包括公共区域、患者区域、洁净区域、厕所、洗手间等位置。

（2）按需配置保洁人员，及时清理垃圾，切勿堆积。

（3）如人手不足，及时调派。

（4）对患者进行引导，做好个人卫生。

5.1.2.3　具体要求

（1）每日9：00、14：00、19：00（三餐后）定时做好卫生清理，发现不符合规定的立即整改，要求区域保洁人员每隔1小时在负责区域进行签到。

（2）垃圾清运：在各层走道、门口、餐饮区等配备垃圾桶，由保洁人员及时清理垃圾至垃圾房，切勿堆积。

5.1.3 防护医用物资保障

5.1.3.1 人员配备

负责人2人，提供联系方式，与其他各部门做好协调。

5.1.3.2 具体工作

根据每日提交医生排班、护士排班，制作名单，保证医务人员的防护物资；根据临时工作安排，做好临时防护物资使用登记；为节省防护用品，领取时限6小时起算；做好防护服领取账目，保证账目准确，账物相符；设立专门防护物资发放岗，24小时轮班；随时供应患者区域氧气瓶、药品等。

5.1.3.3 具体要求

（1）防护物资库存不足100套时（具体标准根据方舱内医务人员数量而定），及时上报协调。

（2）在物资发放处，做好防护物资穿脱流程说明，避免交叉感染。

（3）物资发放处，领取人根据排班，报姓名、实名签字领取。

5.1.4 生活物资保障

5.1.4.1 人员配备

负责人2人，提供联系方式，与其他各部门做好协调。

5.1.4.2 工作内容

（1）为患者提供充足的日常生活用品，如被子、电热毯、杯、盆、毛巾等。

（2）及时协调水、电、网等供应，保证患者有热水、有电源可用。

（3）做好工作区域布置，包括电脑、文具、桌椅等办公用品。

（4）与捐赠处对接，做好捐赠物资入库、发放，及时让捐赠物品物尽其用。

5.1.4.3　具体要求

（1）根据患者床号，做好领取登记清单。

（2）每日统计物资发放情况，保证物品充足。

5.1.5　药品管理

5.1.5.1　制定方舱庇护医院药品目录

根据方舱庇护医院收治的患者特点，结合相关诊疗方案与指南，综合临床一线专家意见，对方舱庇护医院内COVID-19治疗药品需求进行评估分析，以对症治疗、并发症防治、基础疾病治疗及急救药物的供应为主，由药学部临床药师和采购部门最终确定方舱庇护医院药品目录。后期根据临床实际情况，定期对药品数量、品种进行调整、补充。

方舱庇护医院药品目录包括以下类别的药物：

（1）抗病毒药、抗菌药、解热镇痛药、止咳平喘化痰药、胃肠用药；

（2）镇静安眠药；

（3）降压、降糖、调脂等慢病用药；

（4）中成药或其他经临床验证COVID-19患者康复有效的民族医药；

（5）急救药物。

5.1.5.2　建立方舱庇护医院药房

以卓尔（江汉武展）方舱医院为例。卓尔（江汉武展）方舱医院药房占地约30平方米，设置在洁净区，并划分为合格区、不合格区及二级库区。配备计算机、打印机及消防、防盗等基本设施，保证药品存储条件符合相关管理规定的要求。

5.1.5.3　药品采购供应

药品由药品保障供应组负责，制订采购和请领计划，统一采购。药事应急领导小组按照制定的方舱庇护医院药房药品目录，确定紧急采购的药品名单。在保障临床基本药物、常用药物供给的基础上，重点保障应对COVID-19防治相关药物的供给，并划定专区储存。药品采购必须从有合法资质的药品经营企业购入，将有关业务关系的经营企业和业务人员的资质备案。发生药品短缺时应积极与供货商协商沟通，督促其扩大进货渠道或以区域间调货等方式予以解决；存在采购困难的，应积极寻找替代药品，并向临床发布药品指引。

5.1.5.4　捐赠药品管理

（1）接受捐赠药品的标准

1）境内生产的药品，必须是经当地药品监管部门批准生产、获得批准文号且符合质量标准的品种，有效期限距失效日期须在6个月以上。

2）境外生产的药品，应是当地药品监管部门批准注册的品种，以及国际上通用药典收载、在注册国合法生产并上市且符合质量标准的品种，有效期限距失效日期须在12个月以上；药品批准有效期为12个月及以下的，有效期限距失效日期须在6个月以上。

（2）接受捐赠药品的标准流程

1）制订方舱庇护医院捐赠药品目录：讨论制定接受捐赠药品目录，并定期更新，目录内的药品可直接接受捐赠。

2）非目录内捐赠药品处理流程：药品保障供应组接到捐赠方捐赠意愿后，将捐赠方提供的产品资料、说明书等交临床药学进行临床论证及评估，或征询临床医疗团队和临床专家意见，确定是否接受捐赠。

3）接受捐赠药品的数量：依据药品用法用量、疫情时间、疫情发展阶段等方面酌情考虑，由药事应急领导小组确定。

5.2　文化保障

方舱庇护医院实际上是轻症患者的“特殊社区”。为了丰富方舱庇护医院患者的文化生活，增强他们战胜病毒的信心和勇气，建议在方舱庇护医院内设置：

（1）图书角：每个方舱设一个书架，摆放优秀图书免费供患者阅读（图5-2）。

图5-2　卓尔（汉口北）方舱医院图书角

（2）食品角：每个方舱可设一个爱心食品角，摆放一些爱心食品，如方便面、牛奶和水果，为患者免费供应（图5-3）。

图5-3　卓尔（武汉客厅）方舱医院食品角

（3）充电站：每个方舱设一至多个免费充电站，方便患者充电（图5-4）。

图5-4　卓尔（武汉客厅）方舱医院充电站

（4）娱乐角：在每个方舱空旷的地方安装1～2台电视机，一方面供患者收看各类电视节目；另一方面配合组织广场舞、游戏节目、诗歌朗诵、合唱等文艺节目，丰富患者的生活，增强他们战胜病毒的信心和勇气（图5-5、图5-6）。

图5-5　卓尔（武汉客厅）方舱医院娱乐角

图5-6　卓尔（武汉客厅）方舱医院娱乐角的广场舞

（5）心理辅导：新冠肺炎是一种全新的传染疾病，突然来袭，往往会给患者带来一系列应急障碍与焦虑，需要进行一定的心理健康疏导。

5.3 安全保障

5.3.1 方舱庇护医院医疗废弃物管理制度

（1）切实落实主体责任。高度重视舱内产生医疗废物管理，切实落实主体责任，各分区负责人为医疗废物管理的第一责任人，产生医疗废物的具体操作人员是直接责任人。加大环境卫生整治力度，及时处理产生的医疗废物，避免各种废弃物堆积，努力创造健康卫生环境。

（2）培训制度。所有工作人员包括医生、护士、技术人员、管理人员和工人上岗前接受感染控制部门的统一培训。

（3）督导制度。由感染控制部门负责对各区域医疗废物的收集和处理进行定时巡查和问题收集、反馈及整改督导。

（4）分类收集制度。舱内各区产生的所有废弃物，包括医疗废物和生活垃圾，均应当按照医疗废物进行分类收集。

（5）规范包装容器。医疗废物专用包装袋、利器盒的外表面应当有警示标识，在盛装医疗废物前，应当进行认真检查，确保其无破损、无渗漏。医疗废物收集桶应为脚踏式并带盖。医疗废物达到包装袋或者利器盒的3/4时，应当有效封口，确保封口严密。应当使用双层包装袋盛装医疗废物，采用鹅颈结式封口，分层封扎。

（6）做好安全收集。按照医疗废物类别及时分类收集，确保人员安全，控制感染风险。盛装医疗废物的包装袋和利器盒的外表面被感染性废物污染时，应当增加一层包装袋。分类收集使用后的一次性隔离衣、防护服等物品时，严禁挤压。每个包装袋、利器盒应当粘贴中文标签，标签内容包括医疗废物产生单位、产生部门、产生日期、类别，并在特别说明中标注“新型冠状病毒感染肺炎”或者简写为“新冠肺炎”。

（7）潜在污染区和污染区产生的医疗废物处置。在离开污染区前应当对包装袋表面采用1000mg/L的含氯消毒液喷洒消毒（注意喷洒均匀）或在其外面加套一层医疗废物包装袋；清洁区产生的医疗废物按照常规的医疗废物处置。

（8）做好病原标本处理。医疗废物中含病原标本和相关保存液等高危险废物，应当在产生地进行压力蒸汽灭菌或化学消毒处理，然后按照感染性废物收集处理。

（9）医疗废物的运送贮存

1）安全运送管理。在运送医疗废物前，应当检查包装袋或者利器盒的标识、标签以及封口是否符合要求。工作人员在运送医疗废物时，应当防止造成医疗废物专用包装袋和利器盒的破损，防止医疗废物直接接触身体，避免医疗废物泄漏和扩散。每天运送结束后，对运送工具进行清洁和消毒，含氯消毒液浓度为1000mg/L；运送工具被感染性医疗废物污染时，应当及时消毒处理。

2）规范贮存交接。医疗废物暂存处应当有严密的封闭措施，设有工作人员进行管理，防止非工作人员接触医疗废物。医疗废物宜在暂存处单独设置区域存放，尽快交由医疗废物处置单位进行处置。用1000mg/L的含氯消毒液对医疗废物暂存处地面进行消毒，每天两次。医疗废物产生部门、运送人员、暂存处工作人员以及医疗废物处置单位转运人员之间，要逐层登记交接，并说明其来源于新型冠状病毒感染的肺炎患者或疑似患者。

3）做好转移登记。严格执行危险废物转移联单管理，对医疗废物进行登记。登记内容包括医疗废物的来源、种类、重量或者数量、交接时间、最终去向以及经办人签名，特别注明“新型冠状病毒感染肺炎”或“新冠肺炎”，登记资料保存3年。要及时通知医疗废物处置单位进行上门收取或自建医疗废物处理点，并做好相应记录。各级卫生健康行政部门和方舱庇护医院要加强与生态环境部门、医疗废物处置单位的信息互通，配合做好新冠肺炎疫情期间医疗废物的规范处置。

5.3.2　方舱庇护医院院感管理方案

5.3.2.1　工作目标

为降低新冠肺炎在方舱庇护医院内的传播风险，规范医务人员在内的所有方舱庇护医院工作人员行为，避免感染。

5.3.2.2　组织架构

设置医院感染控制委员会，由医院院长、医疗副院长、护理副院长、院感副院长、后勤副院长、病区护士长、病区行政主任组成。院感副院长牵头设立院感工作小组，包括病区护士长、院感医生、院感护士及后勤部门联络人负责日常院感防控工作。

5.3.2.3　工作内容

（1）区域划分。明确界定院内污染区、半污染区、清洁区区域，并在院内醒目位置张贴公示，各区交界处设置醒目警示标志，污染区出入口必须安排专人监督检查人员出入，确保符合院感规范。

（2）开展全员培训。全体人员严格执行培训后上岗制度。依据岗位性质及工作特点，确定不同人员的培训内容，所有需要进入污染区人员需重点培训，熟练掌握新冠肺炎防控知识、方法与技能，提高防控意识。开展所有工勤人员的培训，协助做好环境清洁消毒、患者转运、医疗废弃物处置等。

（3）医务人员及工勤人员防护。在严格落实标准预防的基础上，强化接触传播、飞沫传播和空气传播的感染防控。正确选择和穿脱防护用品，严格执行手卫生。

5.3.2.4　防护制度

（1）个人防护分级制度（表5-1）

表5-1　方舱庇护医院个人防护分级制度表

防护内容	一级防护	二级防护	三级防护
帽子	*	*	*
隔离衣	*		
防护服		*	*
一次性医用外科口罩	*		
医用防护口罩		*	*
护目镜/防护面屏		二选一	都要
手套	*	*	*
长筒靴/保护性鞋套		*	*

（2）不同工作区域的防护等级（表5-2）

表5-2　方舱庇护医院不同工作区域防护等级表

工作区域/工作内容	一级防护	二级防护	三级防护污染区
半污染区		*	
清洁区	*		
标本采集（呼吸道标本）			*
标本采集（非呼吸道标本）		*	
标本运送		*	
消毒（污染区及半污染区）		*	
消毒（清洁区）	*		
患者护送及转运		*	
核酸检测			*
实验室检查（非呼吸道标本）		*	
影像学检查		*	

（3）脱防护服规范（图5-7）

（4）环境清洁消毒

严格执行《医疗机构消毒技术规范》，做好诊疗环境（空气、物体表面、地面等）、医疗器械、患者用物等的清洁消毒，严格患者呼吸道分泌物、排泄物、呕吐物的处理，严格终末消毒。污染区、半污染区、清洁区的消毒保洁用具使用不同颜色标示，不得混用。

（5）患者管理及教育

积极开展方舱内患者的教育，指导个人防护及咳嗽礼仪。转运或出舱进行辅助检查、影像学检查途中全程应佩戴口罩。

脱防护服标准操作规程（SOP）

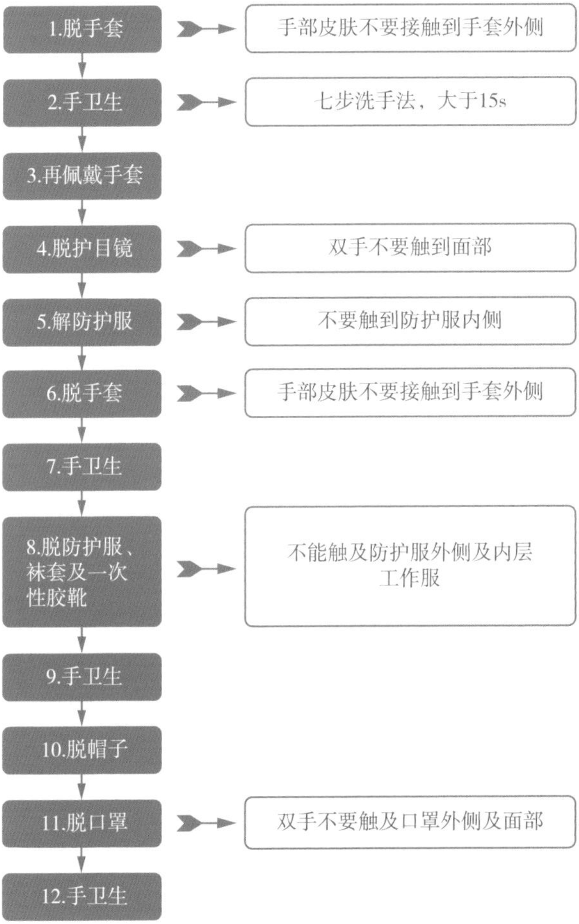

1.脱手套 → 手部皮肤不要接触到手套外侧
2.手卫生 → 七步洗手法，大于15s
3.再佩戴手套
4.脱护目镜 → 双手不要触到面部
5.解防护服 → 不要触到防护服内侧
6.脱手套 → 手部皮肤不要接触到手套外侧
7.手卫生
8.脱防护服、袜套及一次性胶靴 → 不能触及防护服外侧及内层工作服
9.手卫生
10.脱帽子
11.脱口罩 → 双手不要触及口罩外侧及面部
12.手卫生

图5-7　方舱庇护医院脱防护服标准操作规程

5.3.3　安保保障

5.3.3.1　人员配备

负责人1人，提供联系方式，与其他各部门做好协调。

5.3.3.2　具体工作

（1）严禁闲杂人等随意进出。

（2）发现骚乱、争端及时制止劝阻。

（3）协调帮助各种物资搬运。

5.3.3.3　具体要求

（1）进出入口，至少2名安保看管。

（2）患者区域，24小时保安巡视。

5.3.4　方舱庇护医院突发应急事件预案

5.3.4.1　突发纠纷事件应急预案

（1）工作目标：防止舱内人员因情绪过激出现危害他人的情况；用于方舱庇护医院的突发纠纷或暴力事件应对工作。

（2）应急程序（图5-8）

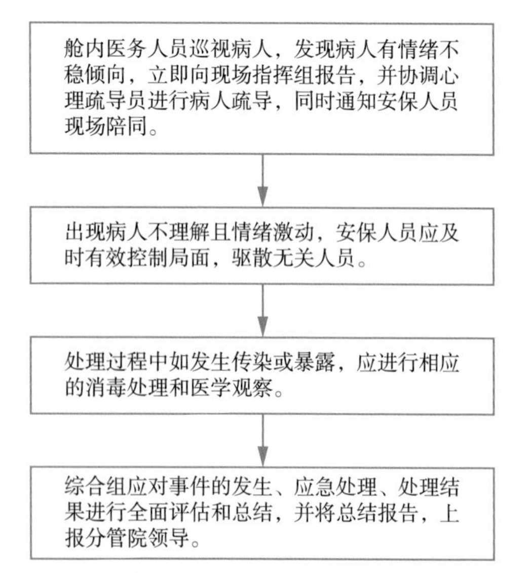

图5-8　方舱庇护医院应急处理程序

5.3.4.2　停水停电应急预案

工作目标：应对停水停电突发事件，预防和控制停水停电损害，保障正常诊疗秩序，确保患者及方舱医务工作人员的生命财产安全，维护医院安全稳定。应急程序：

（1）信息报告：发生停水停电，第一发现人、各接报人必须立即按应急预案规定流程和要求上报。

（2）先期处置：突发事件发生后，各接报人在完成信息报告的同时，要进行先期处置，或根据职责和规定的权限启动现场处置方案或相关应急预案，及时、有效地进行应对控制事态。

（3）应急响应：对于先期处置未能有效控制事态的，要及时启动停水停电专项应急预案，由相关责任人统一指挥或指导有关部门开展处置工作。医疗救治组组长负责指挥应急救治。医务人员应按照科室停水停电现场处置方案组织本片区的救治工作。

另外，水电供应应急保障：

（1）后勤工作组成员应随时保证通信联系方式畅通；

（2）日常做好供水管路、电路、开关阀门的检查，发现问题及时处理；

（3）对相关后勤人员进行培训，要求都能知晓应急供水、管路布局及应急操作流程；

（4）水电维修人员实行24小时值班制度，做到24小时随叫随到。

5.3.4.3　消防事件应急预案

工作目标：为了保障医院区域内人员和财物安全，一旦发生火灾，在统一指挥下，各司其职，恪尽职守，最大限度地避免或减轻人员伤亡和财产损失；使方舱内人员掌握必要的逃生技能。一旦遇到紧急情况，能够及时利用安全疏散通道迅速撤离到就近应急避难场所，学会自救和救人。

应急程序：

（1）发现火灾后，现场负责人员立即组织值班人员，使用就近的灭火设备进行灭火，灭火过程中应根据火灾性质（如发生电气火灾时，必须尽快将故障部位相关电源开关切断），并通知安保人员确认火灾具体情况并拨打119报警。

（2）疏散人群，通过应急广播向火灾现场发出疏散指令，由各区域值班医护人员引导各区域患者有序撤离火灾现场，疏散引导组工作人员要分工明确，统一指挥。

（3）通知就近酒店休息的医务人员在现场及时救治火场受伤人员，必要时与其他近邻医院联系救治工作。

（4）安保人员迅速赶赴火场，进行现场警戒，维持秩序。

（5）后勤组人员对被抢救、转移的物资进行登记、保管，对火灾损失情况协同有关部门进行清理登记。

5.3.5 通风系统运维保障

（1）根据污染区、半污染区、清洁区、医务人员通道、患者通道的划分，计算设定的运行方案，开启或关闭部分新风阀、送风阀、排风阀，调整部分风阀的角度，开启或关闭风口或窗，开启或停用部分通风与空调设备。

（2）随时监测送、排风机故障报警信号，保证风机正常运行；随时监测送、排风系统的各级空气过滤器的压差报警，及时更换堵塞的空气过滤器，保证送、排风的风量。

（3）空气处理机组、新风机组应定期检查，保持清洁。

（4）新风机组粗效滤网宜每2天清洁一次；粗效过滤器宜1～2月更换一次；中效过滤器宜每周检查，3个月更换一次；

亚高效过滤器宜每年更换，发现污染和堵塞及时更换；末端高效过滤器宜每年检查一次，当阻力超过设计初阻力160Pa或已经使用3年以上时宜更换。

（5）排风机组中的中效过滤器宜每年更换，发现污染和堵塞及时更换。

（6）定期检查回风口过滤网，宜每周清洁一次。如遇特殊污染，及时更换，并用消毒剂擦拭回风口内表面。

（7）设专门维护管理人员，遵循设备的使用说明进行保养与维护；并制定运行手册，有检查和记录。

（8）排风高效空气过滤器更换操作人员须做好自我防护，拆除的排风高效过滤器应当由专业人员进行原位消毒后，装入安全容器内进行消毒灭菌，随医疗废弃物一起处理。

5.4　志愿者服务

借鉴其他应急志愿者服务的经验，新冠肺炎作为突发性公共卫生事件，在政府或专业资源不足的情况下，需快速组织调配社会资源，尤其是志愿者参与新冠肺炎疫情防控后勤保障工作，其在自身条件许可的情况下，自愿为疫情防控服务而不获取任何利益的社会团体或个人。

5.4.1　条件与要求

（1）参与疫情防控要严格按照疫情防控机构整体安排，在方舱庇护医院统一组织指挥和调度下，有序开展疫情防控志愿服务。

（2）志愿者服务人员必须身体健康，并有必要的防护措施和装备，接受一定的服务培训，且合理确定服务半径，不跨区域开展志愿服务。

（3）应优先录用有一定医学、心理学专业的志愿者参与工作，在疫情防护宣传、政策措施解读、稳定患者心理情绪等方面提供具有一定专业水平的志愿服务。

（4）志愿者服务要坚持“安全第一”原则，强化疫情防控志愿者的管理和培训，严格落实对在岗志愿者的防护措施，坚决做到防护措施不到位的绝不上岗、防护培训不合格的决不上岗。科学设置志愿者岗位，严格控制志愿者数量，合理设置志愿者服务工作时长。

5.4.2　服务分类

专业辅助服务类：这类志愿者需要进舱或其他特殊指定的高风险区域，与一线医护人员共同维持方舱庇护医院的常规治疗服务。

医疗物资服务类：按分工为方舱庇护医院配送、分发医用防护物资，提供药用和医疗器材等。并协助做好物资的统计和专业存放。

后勤保障服务类：为一线医护人员和方舱里的患者提供各种必要的生活服务，包括车辆通勤服务，提供餐饮、爱心物资等，同时保障舱内水、电、暖气等24小时供应。

5.4.3　服务内容

（1）专业辅助服务。主要是维护方舱内公共卫生秩序，协调调度急需医疗物资和患者收治、分诊、转诊等信息，打扫舱内清洁卫生。比如，在卓尔（武汉客厅）方舱医院就有54名清洁工志愿者，每天三班倒，24小时清洁方舱内的污染物，并严格按规定消杀处理。

（2）医疗物资保障服务。运输、配送并统计各类医疗物资，包括药品、防护服、呼吸机、口罩和手套等，以便医务人员安心、放心为舱内患者提供不间断治疗服务。

（3）生活后勤保障服务。为一线医护人员和方舱里的患者提供包括但不限于餐饮、车辆通勤、爱心物资等方面的物资保障和运动文化娱乐方面的文化服务；同时为方舱庇护医院内的供电、供水、供暖设施的正常运转提供不间断保障，并应对突发事件发生。例如，卓尔（武汉客厅）方舱医院从搭建到最后休舱，卓尔公益基金会100多名志愿者在72小时内完成搭建1500张床具，帮助设立图书角、充电角、食品角、电视角等功能区域，并配送爱心物资或分发社会捐赠的物资，24小时维护供水供电供暖等设施的正常运转，确保零故障。

参考文献

[1] Chen S，Zhang Z，Yang J，et al. Fangcang shelter hospitals：a novel concept for responding to public health emergencies [J]. The Lancet，2020, 395：1305-1314.

[2] 国家卫生健康委医政医管局，国家卫生健康委医疗管理服务指导中心. 方舱医院工作手册（第三版）[S]. 2020.

[3] 浙江省住房和城乡建设厅. 方舱式集中收治临时医院技术导则（试行）[S]. 2020.

[4] 湖北省住房和城乡建设厅. 方舱医院设计和改建的有关技术要求（修订版）[S]. 2020.

[5] 住房和城乡建设部办公厅，国家卫生健康委办公厅. 新型冠状病毒肺炎应急救治设施设计导则（试行）[S]. 2020.

[6] 辜明，华小黎，陈骏，等.江汉方舱医院药事管理与药学服务实践[J].中国药师，2020，23（4）：702-706.

鸣　　谢

马云公益基金会

阿里巴巴公益基金会

中国一冶集团有限公司

中信建筑设计研究总院有限公司

湖北省住房和城乡建设厅

浙江省住房和城乡建设厅

中南建筑设计院股份有限公司

武汉地产开发投资集团公司

武汉市汉阳市政建设集团公司

长江建投集团（武汉）产业投资有限公司

武汉市射击运动学校

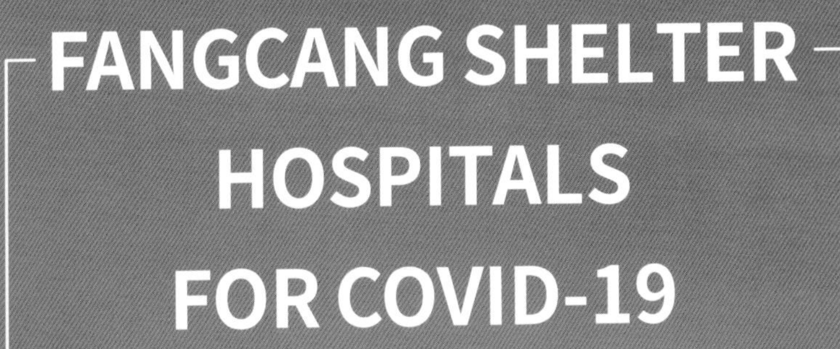

CONSTRUCTION AND OPERATION MANUAL

Yan Zhi Editor-in-Chief
Yan Ge English translator

Editor-in-Chief: Yan Zhi

Contributors: Yan Zhi, Zhang Liang, Du Shuwei,
Fang Li, Pan Zijing, Cheng Longhua

English translator: Yan Ge

English proofreader: Wang Sida

Layout designers: Huang Xuan, Song Jie, Ye Qinyun

Foreword

SARS-CoV-2 is a novel coronavirus that is featured by high infectivity and rapid spread. Transmission of SARS-CoV-2 occurs through droplets and can happen through close personal contact with infected persons. Coronavirus disease 2019 (COVID-19) caused by caused by SARS-CoV-2 has been identified and reported in China and many other countries since late 2019.

As a novel public health concept, the Fangcang Shelter Hospital was first proposed by Dr. Wang Chen, academician of Chinese Academy Engineering & President of Chinese Academy of Medical Sciences and Peking Union Medical College, in Wuhan, China, in February 2020. As the surging number of COVID-19 patients was not able to access to hospital beds and other medical resources, Dr. Wang innovatively proposed that the large-scale public venues such as exhibition centers and indoor stadiums could be converted into Fangcang Shelter Hospitals to receive mild-to-moderate COVID-19 patients with minimum costs within a short period of time. These Fangcang Shelter Hospitals have functions including isolation, triage, basic medical care, close monitoring/rapid referral, and essential living and social engagement, which effectively isolated the infection sources and increased health-care capacity in Wuhan.

Zall Foundation crews had contributed to the design, renovation, and operation of theses shelter hospitals. This booklet is a good summary of their knowledge and experiences, with an attempt to contribute to the global efforts in combating the COVID-19 pandemic.

Yan Zhi

Founder, Zall Foundation

April 2020

Contents

Chapter 1 Proposing the Strategy of Establishing Fangcang Shelter Hospitals

1.1 Background

Wuhan, a metropolitan city with millions of people, went into lockdown on January 23, 2020, to interrupt the transmission of novel coronavirus. The number of confirmed cases of COVID-19 grew rapidly, which overburdened medical resources. A large number of confirmed patients could not be admitted to the hospital and had to choose home quarantine. Prof. Wang Chen, Vice President of Chinese Academy of Engineering, President of Chinese Academy of Medical Sciences & Peking Union Medical College, and an expert in pulmonary and critical care medicine (PC-CM) worked at the frontline to fight against the epidemic and proposed the idea of constructing Fangcang shelter hospital to implement the strategy of "All suspected and con-firmed patients must be admitted to the hospital and all the confirmed patients should be treated" , which was urged by the central government of China. The China Culture Expo Center of Wuhan Salon, Wuhan International Conference and Exhibition Center, and Hongshan Gymnasium in Wuhan were expropriated on February 3 to re-built into the first batches of Fangcang shelter hospitals in Wuhan.

Fangcang shelter hospital was constructed using mainly existing buildings or resources, to admit COVID-19 patients with mild to moderate symptoms in the shortest time and at mini mum cost to the greatest extent. It provided basic medical care for patients, transferred them to designated hospitals, and provides them with basic living conditions; also, it controlled the infection sources effectively and cut off the route of transmission. Thus, it prevented the spread of pandemic, increased the recovery rate, and reduced the case-fatality rate. Although the construction of Fangcang shelter hospital was not a "perfect strategy" , it was feasible and realistic. Fangcang shelter hospital played a key role during COVID-19 prevention and control in China, which may shed light on the global efforts in containing the coronavirus pandemic.

The number of confirmed cases of COVID-19 is on the rise worldwide, which results in a desperate lack of medical resources in all countries, especially the number of ward beds for the treatment of confirmed patients. A significant proportion of confirmed patients cannot be admitted to hospitals and receive appropriate treatment. Self-quarantine at home will put family members at risk, causing cross infection and leading to further spread of the disease. With the establishment of Fangcang shelter hospital severe and critical patients were able to be admitted into designated hospitals, whereas a large number of patients with mild to moderate symptoms were admitted into Fangcang shelter hospital in a centralized manner, which improved the utilization efficiency of medical resources.

1.2 Definition of Fangcang Shelter Hospital

Fangcang shelter hospitals in China were large improvised hospitals that were re-constructed from convention and exhibition centers, stadiums, and other existing large places. It was used to isolate a large number of COVID-19 patients with mild to moderate symptoms from family members and communities and to provide medical care, disease surveillance, and referral as well as living and social spaces for these mild-to-moderate patients.

1.3 Characteristics of Fangcang Shelter Hospital

An article (*Fangcang shelter hospitals: a novel concept for responding to public health emergencies*) focusing on China's construction and use of Fangcang shelter hospitals was published on April 2 in *The Lancet,* one of the world's top medical journals. This article was prepared jointly by Prof. Wang Chen and Institute for Global Public Health of University of Heidelberg, Germany. Three characteristics of Fangcang shelter hospitals are listed in the article: fast construction, large scale, and low cost. These characteristics enable it to efficiently respond to public health emergencies.

Patients unable to be admitted to hospitals were mainly admitted to Fangcang shelter hospital, which not only prevented infection to family members but also provided timely medical treatment for these patients. All patients admitted to Fangcang shelter hospital were tested positive in the nucleic acid testing(NAT). Admitted patients were asked to wear masks and take other preventive measures. Therefore, there was basically no cross-infection among patients admitted into Fangcang shelter hospital.

1.4 Functions of Fangcang Shelter Hospitals

The Fangcang Shelter hospital has five essential functions:

(1) Isolation. For patients with mild to moderate symptoms, it has better effect than self-quarantine at home.

(2) Triage. It provides a strategic function for confirmed COVID-19 patients. Mild to moderate COVID-19 patients are admitted to Fangcang Shelter Hospital for treatment in isolation, while severe and critical COVID-19 patients are treated in normal hospitals, thus effectively releasing capacity pressure of the local hospitals.

(3) Provision of basic medical care. This includes antiviral, antipyretic and antibiotic treatment, support of oxygen and intravenous fluids, and mental health counseling.

(4) Frequent monitoring and rapid referral. The patient's conditions such as breathing frequency, body temperature, and oxygen saturation are measured frequently every day. The patients who meet certain clinical criteria are quickly transferred to designated higher-level medical institutions for treatment. This greatly reduced referral time before the patients get worsened.

(5) Provision of a community for patients with living essentials and social engagement. Fangcang Shelter Hospital provides a community for patients with mild symptoms to moderate symptoms, where mutual assistance between the medical staff and patients and social activities participated by the patients will ease the anxiety caused by the disease and isolation, so as to promote rehabilitation.

Chapter 2 Project Design for Fangcang Shelter Hospitals

2.1 Function Division

Fangcang shelter hospitals shall be designed under the principle of "Three zones, two passages" ; or, the traffic should be organized in a way that medical staff and patients can be separated, and the contaminated passages and the clean passages are separated. The negative-pressure ventilation system should be installed. Moreover, there should be enough space for living and social engagement.

Three areas include contaminated zone, semi-contaminated zone, and clean zone.

- Clean zone: Clean zone is for medical staff members who have low risk of exposure to patients' blood, body fluids, pathogenic microorganisms, and other polluted or infected materials. Confirmed patients with infectious disease are prohibited from entering the area.
- Semi-contaminated zone: Semi-contaminated zone is located between contaminated zone and clean zone, for medical staff members who have medium risk of exposure to patients'

blood, body fluids, pathogenic microorganisms and other contaminated or infected materials.

- Contaminated zone: suspected and confirmed patients with infectious disease are being treated in this area. Blood, body fluids, secretion, medical waste and any contaminated materials are being disposed of in this area.

There should be a clear and obvious sign to distinct different areas, for which isolation belt can be also used. Area can be differentiated by colored sign. "Two passages" are namely health workers' passage and patients' passage. Moreover, cleansing passage and contaminated zone should be strictly separated to avoid any unnecessary interaction between medical staff and patients.

When passing through contaminated zone from clean zone, the entrance used should be labeled as "Sanitary Entrance Room" and "Sanitary Exit Room".

The entering procedure should be: first changing room-second-changing room-buffer room. After all protective equipment is donned, the medical staff can approach to the clean zone.

The returning procedure should be: buffer room-isolation suites taken off-buffer room-showering-changing room, then proceed to clean zone. The sanitary exit room should be distinguished by genders.

2.2 Ward Bed Area Design

The bed area should be divided into male section and female section. Every bed area should be allocated with no more than 42 beds. There should be 2 exits, located within 30m away from any point of the bed area. The corridor between two different areas should be used as evacuation corridors. In a large, open space area, the width of evacuation corridor should be no less than 4m. The evacuation corridor should be indicated with clear sign. The material used for isolation belt located between different bed areas should be incombustible or flam retardant. The surface of the belt should be scrub resistant, while the height of the belt should be no lower than 1.8m. Beds shall be arranged with proper spacing, and comfortable enough for treatment and monitoring. The distance between two parallel beds should be no less than 1.4m. When placing beds in a single line, the distance between bed and the wall opposite should be no less than 1.1m.

2.3 Toilet Design

The toilets for medical and patients should be set up separately. The toilets for patients should be temporary, and a special passage shall be set between the temporary toilet and the ward area. Foam-blocked portable toilets should be used if it is possible. The number of toilet and population ratio should be 1:20 for male, and 1:10 for female. The toilets amount can be adjusted according to reasonable needs and requirements from patients. The polluted water from toilet should be first discharged to disinfection tank first, before distributing into urban sewerage and drainage pipes.

The existed toilets in the public venue can be provided for medical staff and logistics support staff who are in a healthy condition.

Toilet can be closed if it is not in use.

2.4 Firefighting and Barrier-free Design

The admission capacity should be determined based on the width of emergency exit and evacuation stairs. The evacuation safe space should be able to contain 100 people or designed according to relevant fire prevention regulations.

The main entrance and exit, as well as all the channels to different department, should be provided with barrier free corridors. There should be rampway design when there is a height difference; the falling gradient should be strictly followed the relevant constriction code. The width of the barrier free corridor should be large enough to contain both wheelchair patients and the companion medical staff.

2.5 Design of Auxiliary Rooms

There should be a specific locker room for personal belongings, a disinfection and security-check room, and patients changing room located in the entrance. There should be an area for packing and disinfecting when transferring or discharging patients. Moreover, the location of emergency resuscitation room, diet preparation room, equipment storage, filth cleaning room, and daily garbage storage should be near by the bed area. The pharmacy, medical equipment storage, medicine storage, diet preparation room, duty room, and office should be located near the medical staff working area.

2.6 Fangcang Shelter Hospitals Design Cases

Public venues such as indoor stadium, exhibition center, departure hall, factory, and school function hall can be re-developed into Fangcang shelter hospitals. Some reference cases are shown as below.

2.6.1 Single-floor Exhibition Center Fangcang Shelter Hospital

Case: Wuhan Salon: Cultural Exhibition Center under Zall Property was re-developed into a single-floor Fangcang Shelter Hospital in East-west Lake District (See Fig.2-1). Zall (Wuhan Salon) Fangcang Shelter Hospital was reconstructed on the base of three exhibition halls (Halls A, B, and C). The detailed layout is shown in Fig.2-2.

Fig.2-1 Aerial view of Zall (Wuhan Salon) Fangcang Shelter Hospital

We first studied and evaluated the original layout and construction drawing of the exhibition center, then carried out a site investigation. Based on these information and investigation results as well as the relevant regulation and standards, our experts formulated an achievable construction plan.

Zall (Wuhan Salon) Fangcang Shelter Hospital was divided into "three zones and two passages" (See Fig.2-2), and the reconstruction plan was designed according to the function requirements of a professional hospitals. Functional renovation was carried out for the main exhibition hall (Exhibition Hall A) and two side exhibition halls (Exhibition Hall B and Hall C). The contaminated zone was used for medical treatment and the admission of patients; the clean zone was for living engagement and supplies; and the hall in between was used as sanitary room. The medical staff dorm was located near the hospital. The staff members were allowed to return back home after 14 days of quarantine (See Fig.2-3).

Key point 1: Wards and nursing station were designed using the large space in exhibition hall, which are separated and designed in a fishbone layout. The nursing station was located in the center, with paths connecting wards at two sides. Hospital received patient on outboard, and the basic treatment procedures were: admitted from patient's passage in lateral side treatment completed disinfect in clean zone discharge Fig.2-4.

Key point 2: Areas for living engagement and medical and living supplies were provided. The space around the existed entrance for supplies was used as storage for urgent medical supplies. There was also an independent entrance and exit for medical staff, and on-call rooms, offices, consultation rooms and telemedicine room were established nearby. There was a sanitary room that connected the nursing station. (See Fig.2-5 and Fig. 2-6)

2.6.2 Multi-floor Exhibition Center Fangcang Shelter Hospital

Case: W Zall (Jianghan, Wuzhan) Fangcang Shelter Hospital was reconstructed from Wuhan International Exhibition Center in Jianghan District. It was a good example for Multi-floor Exhibition Center Fangcang Shelter Hospital. Just like Zall (Wuhan Salon) Fangcang Shelter Hospital, it was reconstructed based on a large space exhibition hall, except that Zall (Jianghan Wuzhan) Fangcang Shelter Hospital had two floors. The first floor was clean area or semi-contaminated clean area, and the second floor was a contaminated area. (See Fig.2-7 and Fig.2-8)

2.6.3 Indoor Stadium Fangcang Shelter Hospital

Case: The Wuchang Fangcang Shelter Hospital was reconstructed from Hongshan Indoor Stadium in Wuchang, Wuhan. It was redeveloped based on the indoor basketball court. The construction drawings are shown in Figures 2-9, 2-10, 2-11, 2-12, 2-13, and 2-14.

2.6.4 Vacant Factory Fangcang Shelter Hospital

Case: Zall Rehabilitation Station in Changjiang New Town, Wuhan. This Fangcang shelter hospital was a combination of 10 small "shelter rooms" , and the surface area of each shelter room was 12000m^2. The detailed plan is shown in Fig.2-15.

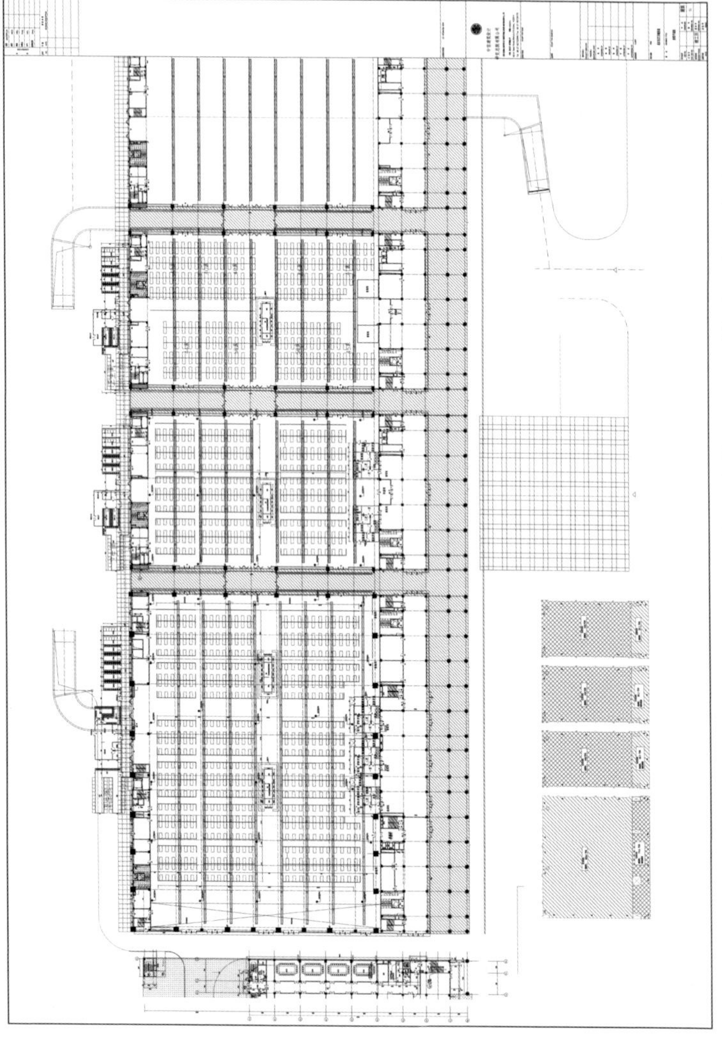

Fig.2-2 Plan of Zall (Wuhan Salon) Fangcang Shelter Hospital

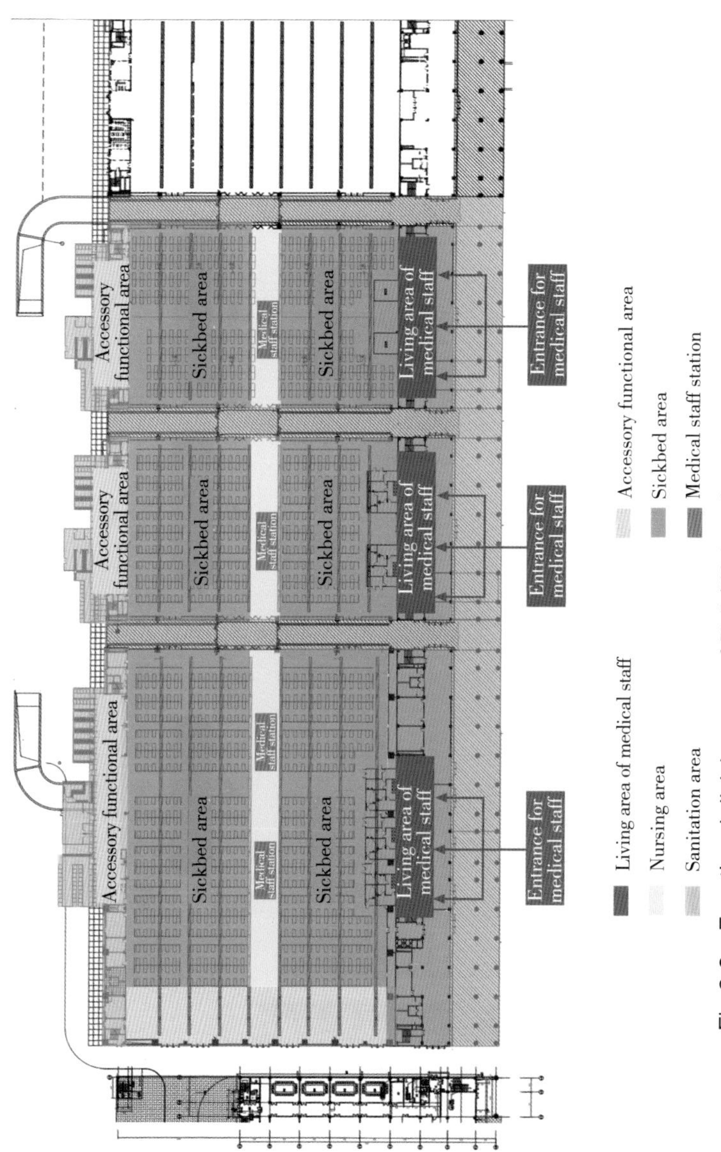

Fig.2-3 Functional division map of Zall (Wuhan Salon) Fangcang Shelter Hospital

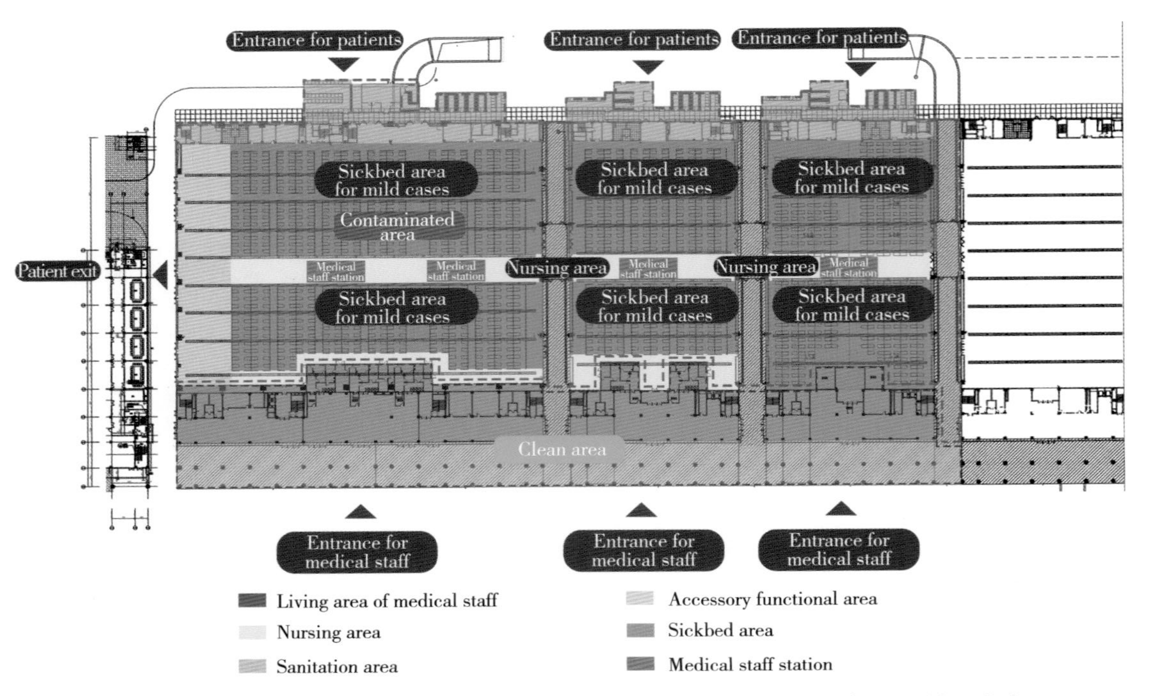

Fig.2-4 Functional division map of Zall (Wuhan Salon) Fangcang Shelter Hospital

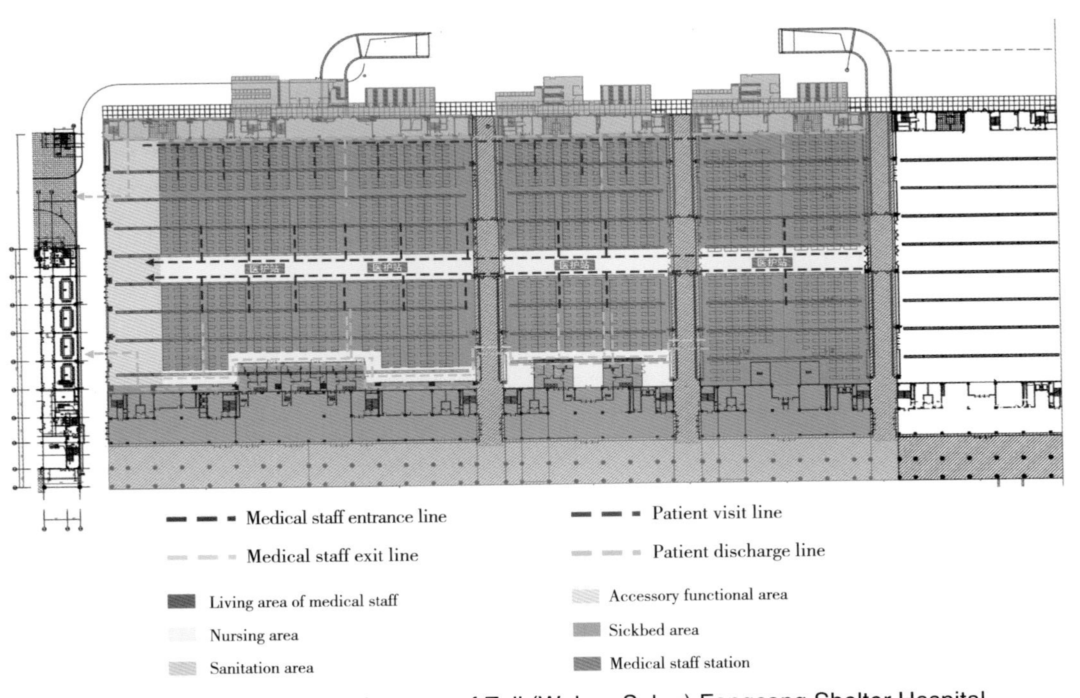

Fig.2-5 Functional division map of Zall (Wuhan Salon) Fangcang Shelter Hospital

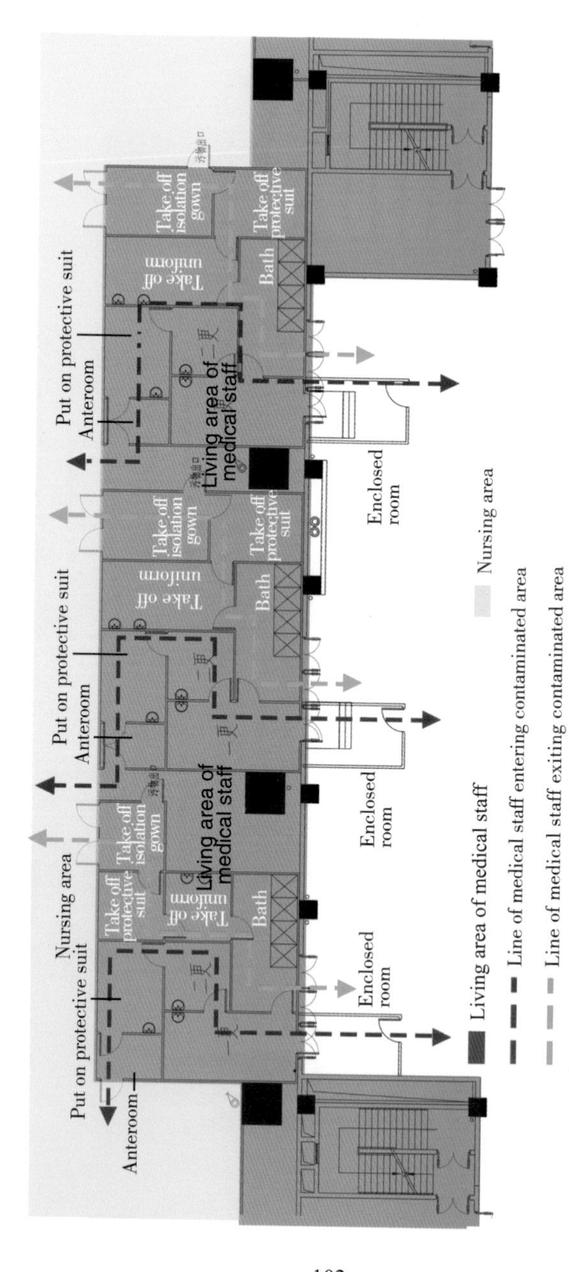

Fig.2-6 Functional division map of Zall (Wuhan Salon) Fangcang Shelter Hospital

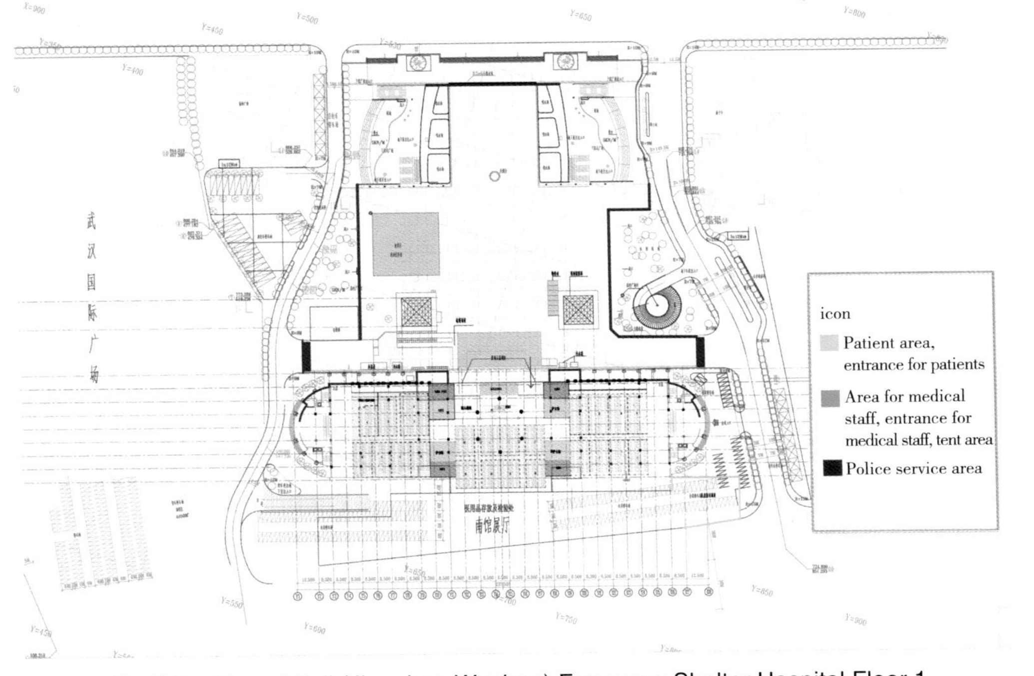

Fig.2-7 Plan of Zall (Jianghan Wuzhan) Fangcang Shelter Hospital Floor 1

Fig.2-8 Plan of Zall (Jianghan Wuzhan) Fangcang Shelter Hospital Floor 2

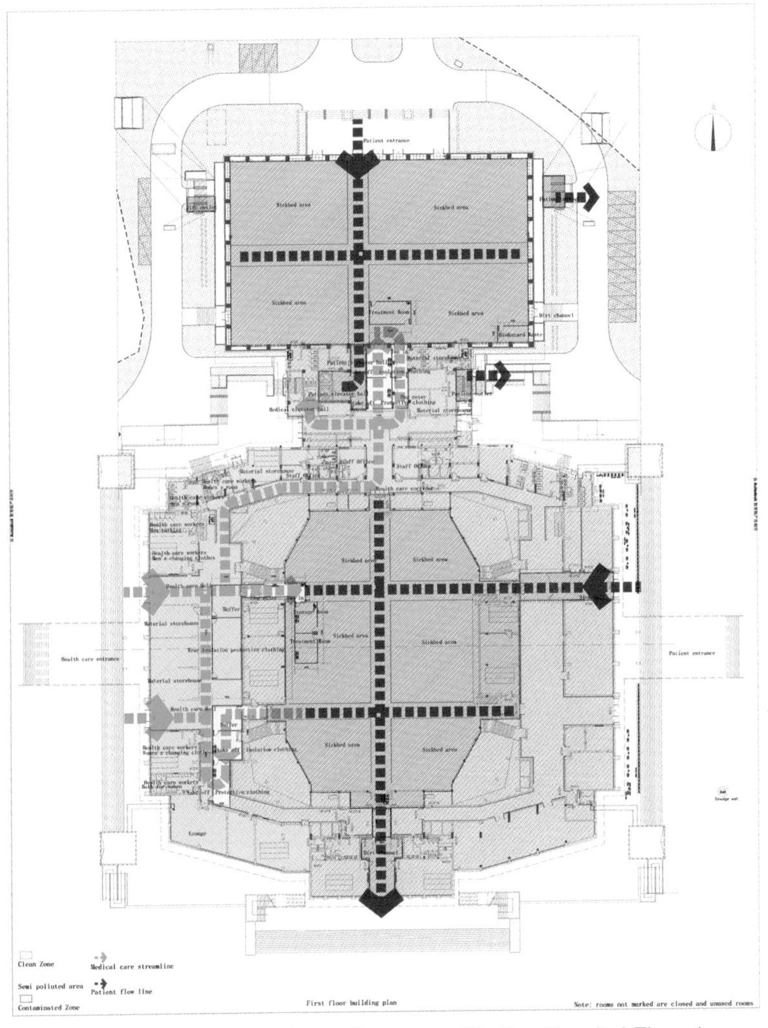

Fig.2-9 Plan of Wuchang Fangcang Shelter Hospital Floor 1

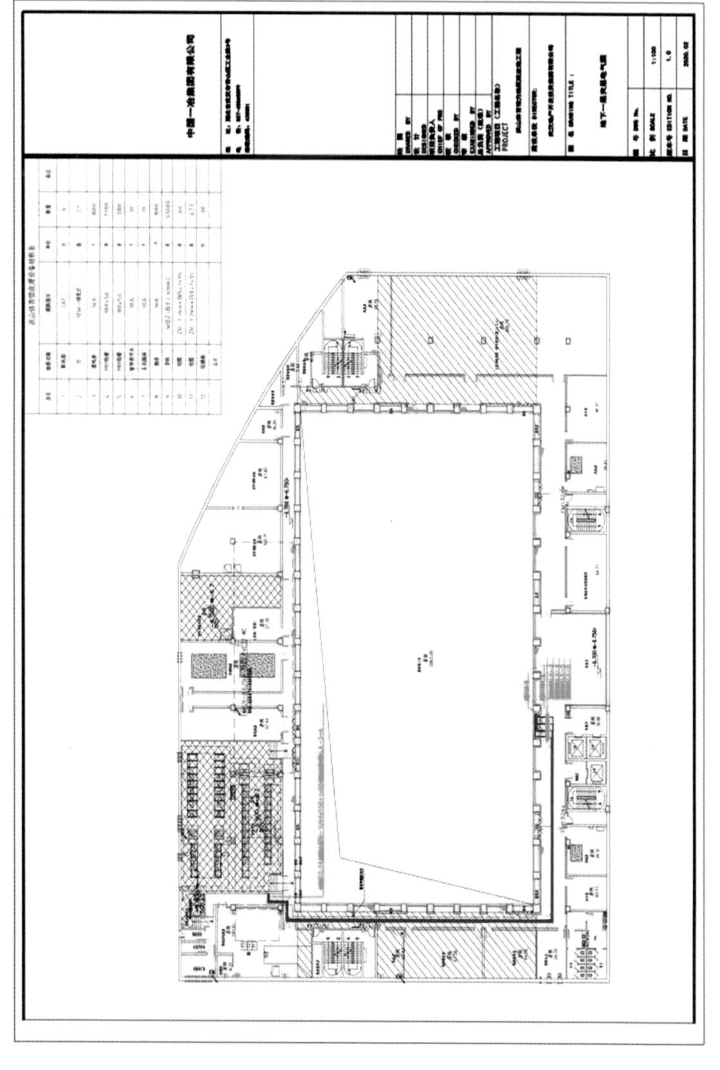

Fig.2-10 Plan of Wuchang Fangcang Shelter Hospital Basement 1 interlayer

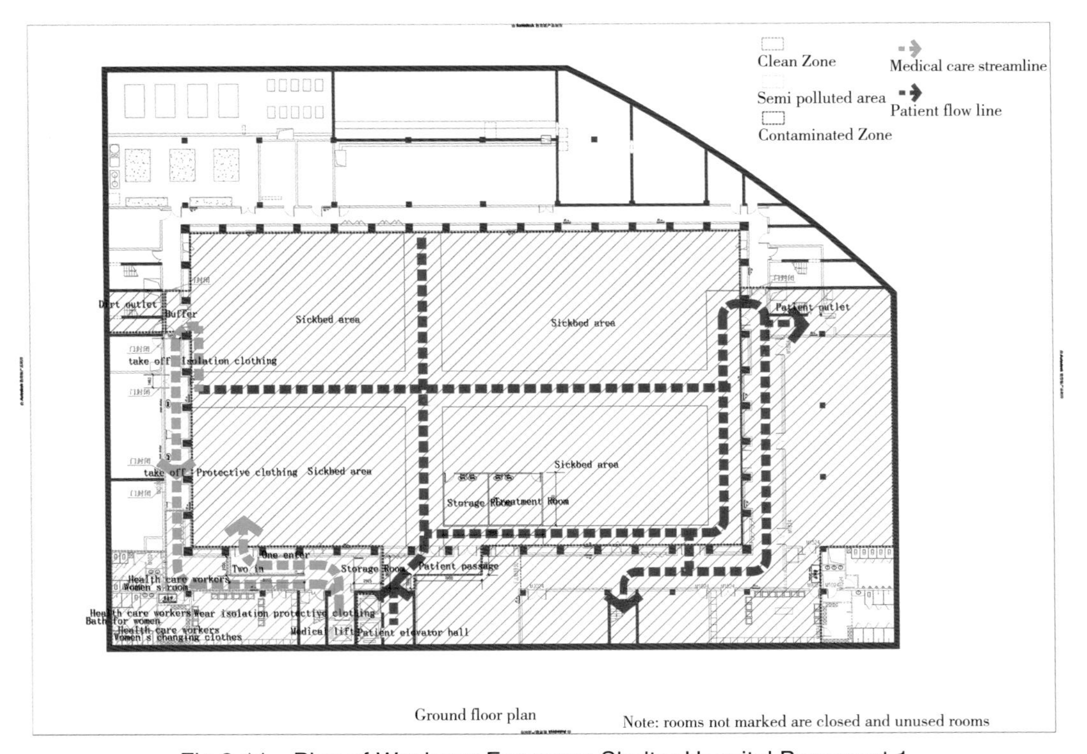

Fig.2-11 Plan of Wuchang Fangcang Shelter Hospital Basement 1

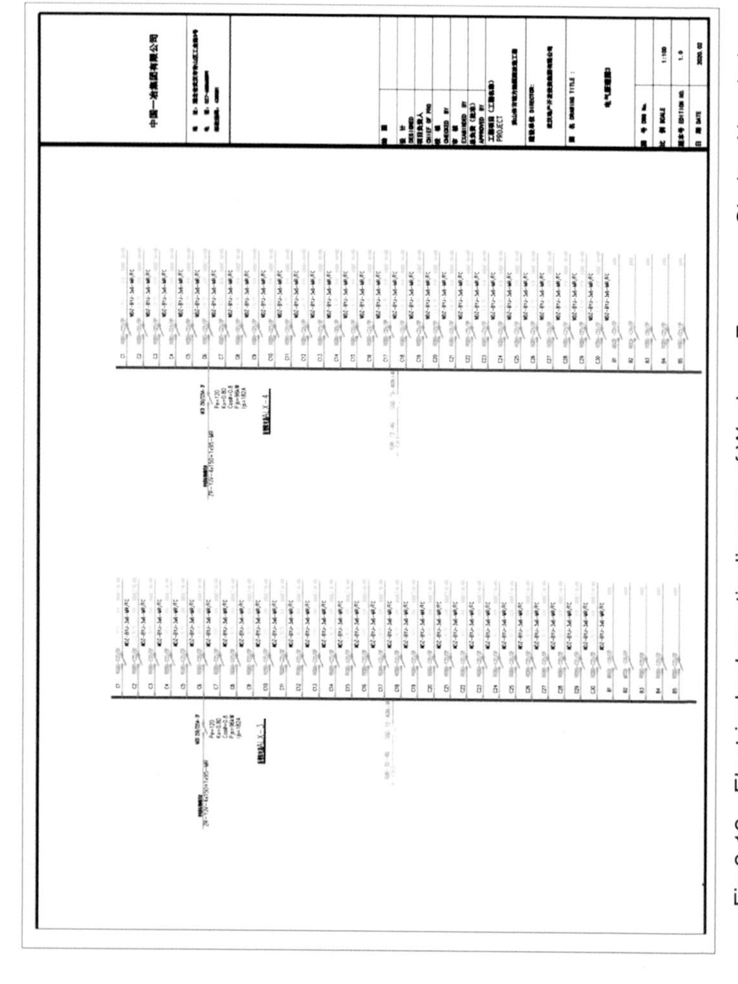

Fig.2-12 Electrical schematic diagram of Wuchang Fangcang Shelter Hospital

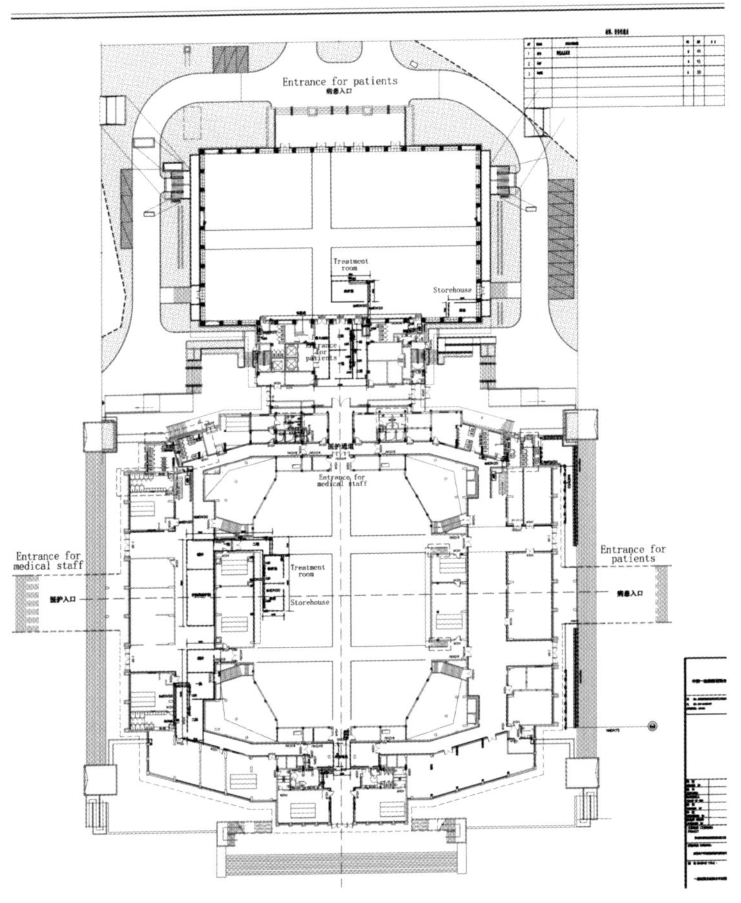

Fig.2-13 Building, water supply, and drainage plan of Wuchang Fangcang Shelter Hospital Floor 1

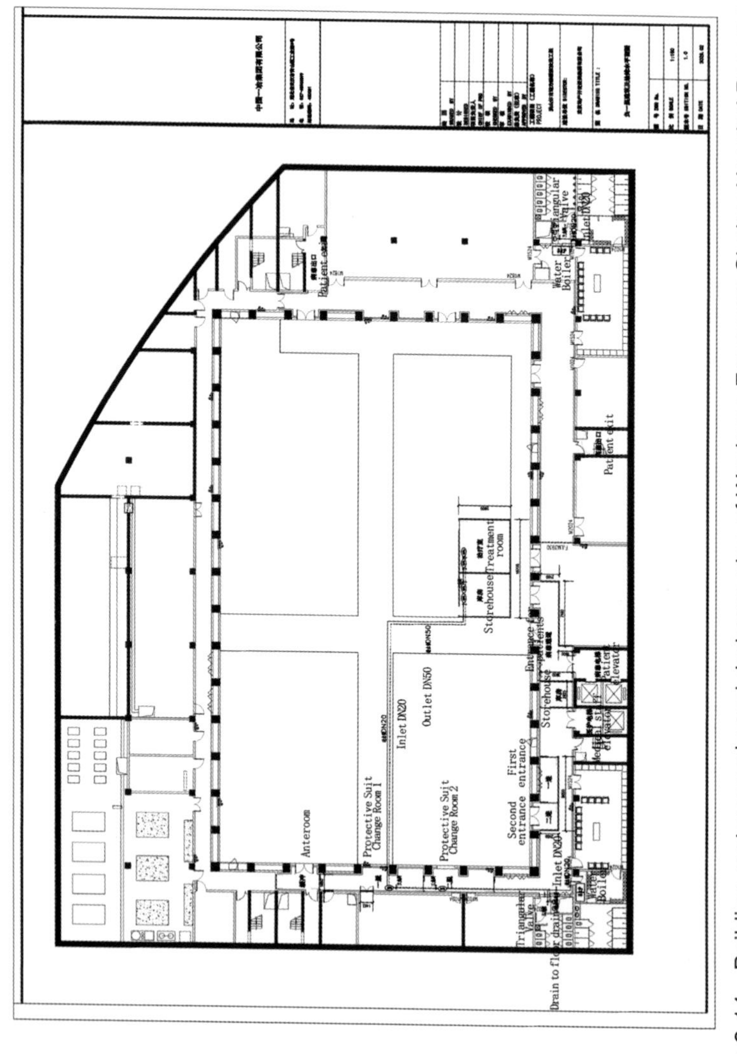

Fig.2-14 Building, water supply and drainage plan of Wuchang Fangcang Shelter Hospital Basement 1

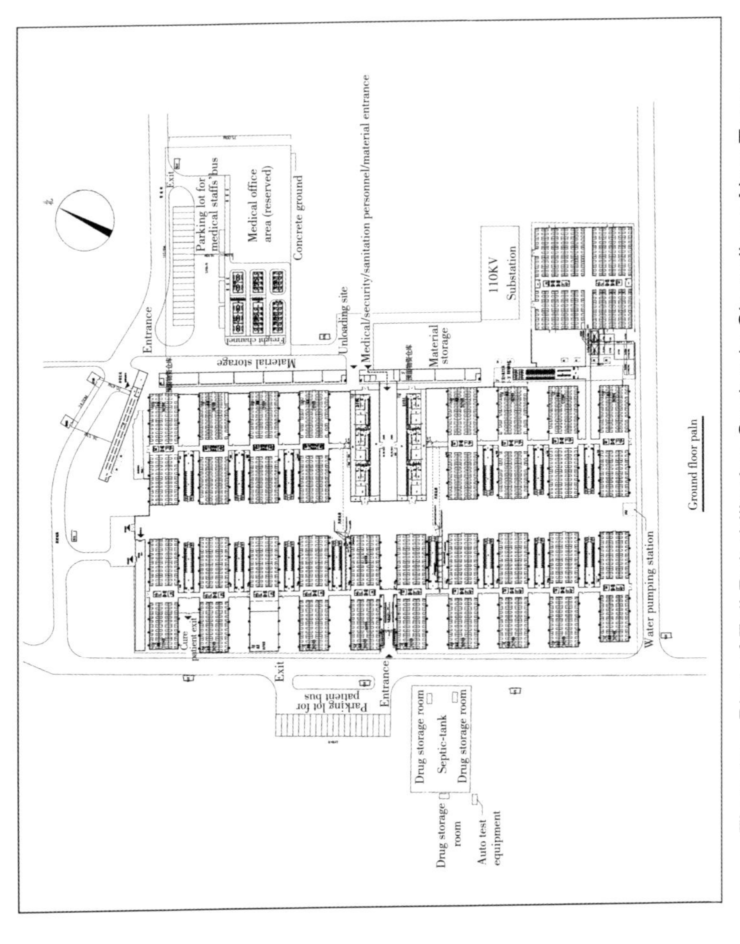

Fig.2-15 Plan of Zall Rehabilitation Station in Changjiang New Town

2.6.5 Departure Hall Fangcang Shelter Hospital

Case: Zall (Hankou North) Fangcang Shelter Hospital. It was reconstructed based on a departure hall in a train station. It was located in Hankou North Passenger Terminal in Huangpi District, Wuhan. The detailed layout plan is shown in Fig.2-16.

2.6.6 Multi-function School Venue Fangcang Shelter Hospital

Case: Hanyang Fangcang Shelter Hospital in Hangyang District, Wuhan. It was reconstructed based on a three-floor multi-function sports hall (surface area: 13,000 m^2) and a tennis court (4,800m^2) in Sports School, Wuhan. The detailed layout plan is shown in Figures 2-17, 2-18, 2-19, and 2-20.

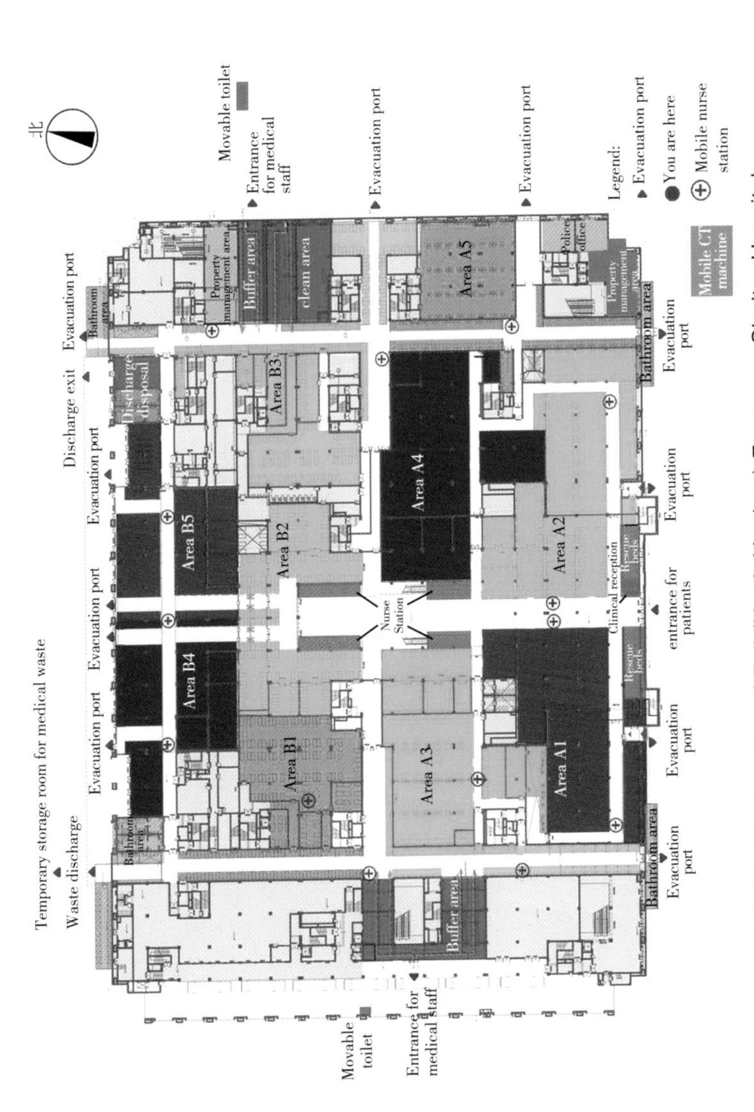

Fig.2-16 Zoning plan of Zall (North Hankou) Fangcang Shelter Hospital

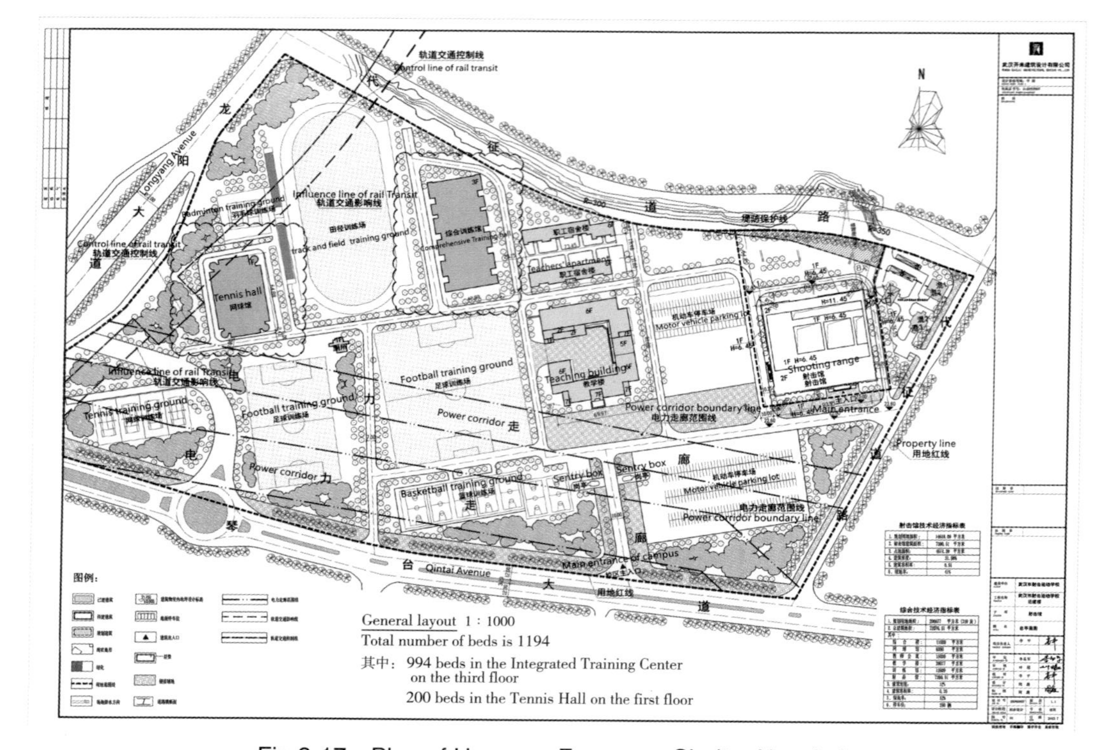

Fig.2-17 Plan of Hanyang Fangcang Shelter Hospital

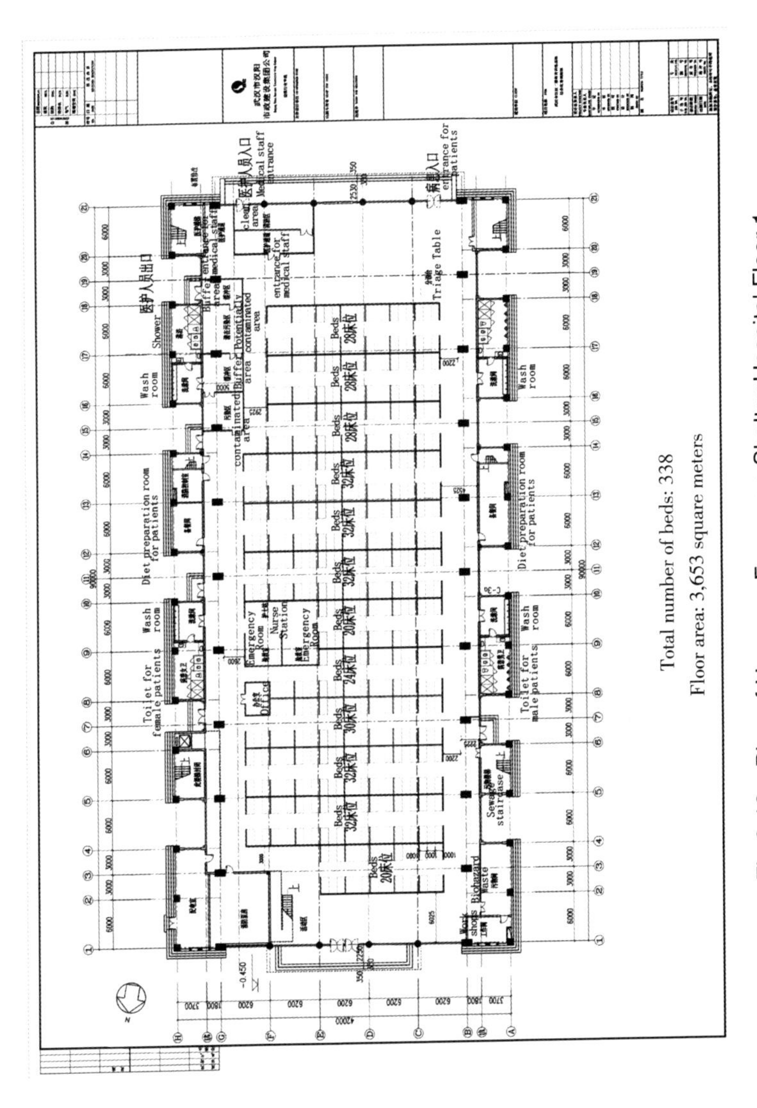

Fig.2-18 Plan of Hanyang Fangcang Shelter Hospital Floor 1

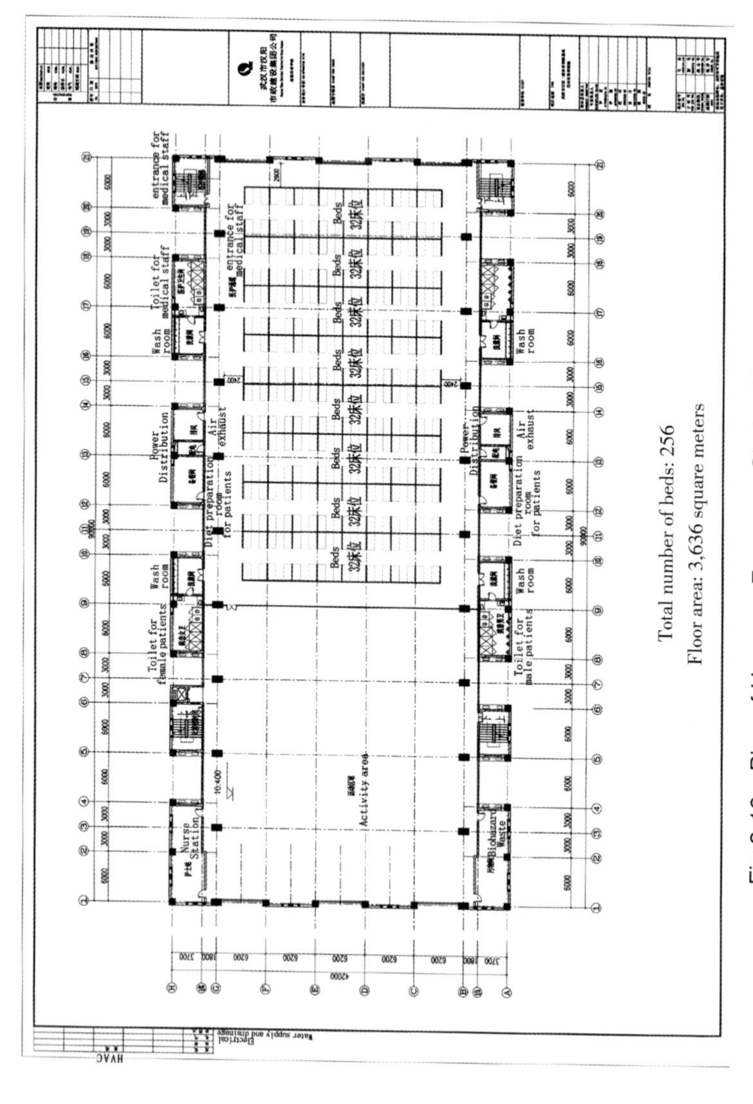

Fig.2-19 Plan of Hanyang Fangcang Shelter Hospital Floor 2

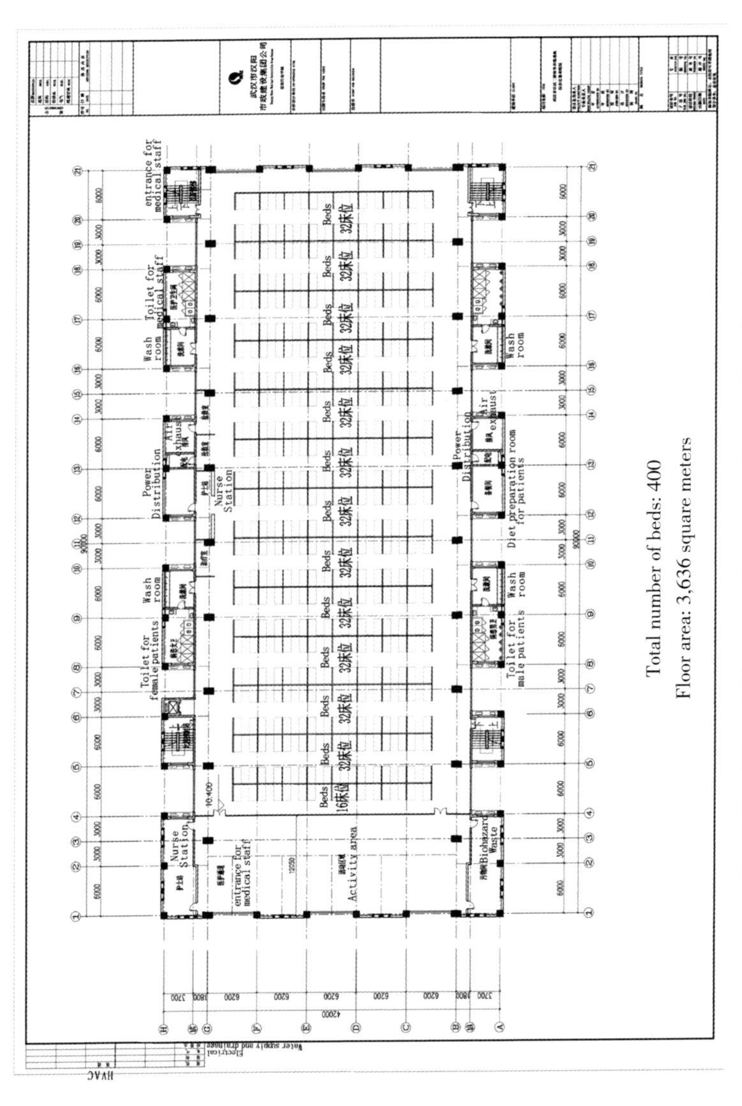

Fig.2-20　Plan of Hanyang Fangcang Shelter Hospital Floor 3

Chapter 3 Fangcang Shelter Hospitals Reconstruction

3.1 Content of Reconstruction

3.1.1 Reconstruction Principles

The reconstruction of Fangcang shelter hospitals involves the following aspects: infrastructure provided by local government; sewage treatment facilities; internal separation belts; indoor equipment and facilities; circulation routes to outside traffic; supplies transportation corridors; protection and improvement of neighborhood environments; epidemic prevention and control facilities; biological safety facilities; and security.

Until the end of expropriation period, the public venues can be solely used as Fangcang shelter hospitals for receiving and treating the suspected and confirmed COVID-19 patients.

The reconstruction of the existed property must strictly comply with the relevant epidemic prevention and control regulations and infectious disease hospitals standards.

If the relevant regulations and standards could not be met,

reasonable adjustment on the property reconstruction plan should be made based on the actual situation.

3.1.2 Cases

Venues such as exhibition hall, indoor stadium, train station departure hall, multi-function sports hall and vacant factory should be renovated in following aspects: sewage system, ventilation system, power system, and ward areas, in order to redevelop into a Fangcang shelter hospital. Zall (Wuhan Salon) Fangcang Shelter Hospital was a good example for Fangcang Shelter Hospital reconstruction project.

(1) Strong electricity wires were installed in halls A, B, and C. Each hall was allocated with approximately 1500 beds. Sockets were installed next to bedside in every ward. For security purpose, there were interior partition devices set up in the halls.

(2) Power systems were installed in nurse station, medical waste room, storage room, treatment room in halls A, B and C.

(3) Mechanical air supply and air exhaust systems in semi-contaminated areas (first-changing room, second-changing room, buffer room, room for donning/doffing PPEs, etc.) in hall A and C were supported by power system, with installation of distribution boxes and electrical cabling nets. There was completed airtight space located in entrances in every semi-contaminated area in halls A and C.

(4) Washrooms and bathrooms should be set up in halls A, B, and C. In hall A, 4 wash rooms were required for collecting filthy cloths, with 40 washing basins and 40 portal electric water heaters. Moreover, two bathrooms were required with 12 sets of showering equipment and 40 portable electric water heaters. Whereas in halls B

and C, 4 wash rooms were required with 40 washing basins and 40 portable electric water heaters. Moreover, 2 bathrooms were required, with 12 sets of showering equipment and 12 portable electric water heaters. Every room should be supported by lightning system, with installation of cabling nets. Power should be provided for disinfection facilities and ventilation system. Distribution boxes and cabling nets should be provided for exterior sewage system.

(5) The exterior water supply system was reconstructed. Main water supply pipe should use DN100 PE, which was connected with branch pipes of water container and matching valves.

(6) The exterior drainage system was reconstructed. Main water drainage pipe should use DN150 UPVC. The entrances of parking basement in halls A and C were placed with 75m^3 glass-made septic tank. The sewage systems were installed in halls A and C, which were complemented with 4 sets of submersible sewage pumps, 2 control panels, and supporting valve equipment.

(7) Mechanical air supply and air exhaust systems were installed in halls A and C, with supporting distribution boxes and cabling nets.

(8) The ward areas in halls A, B and C were separated clearly by belts made in incombustible material, which were artistic and able to protect the privacy of patients.

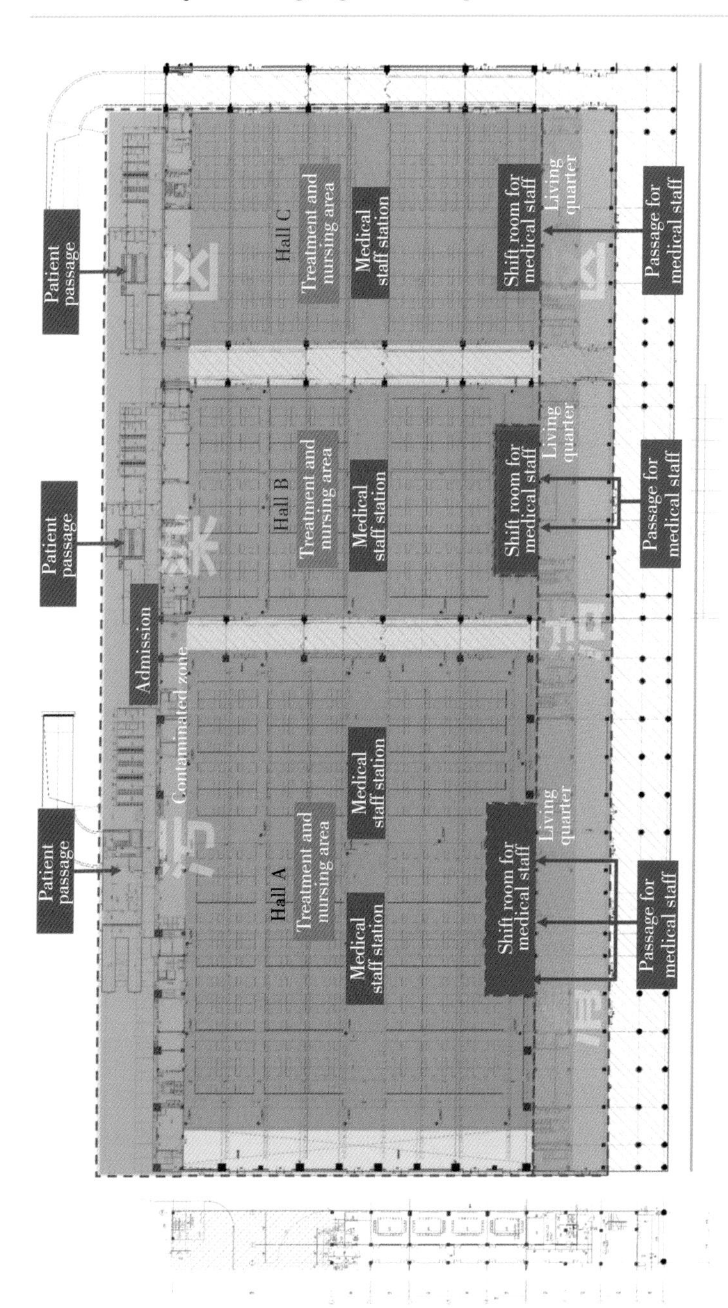

Fig.3-1 Floor plan of Zall (Wuhan Salon) Fangcang shelter hospital

3.2 Requirements on Reconstruction

3.2.1 Site Selection Requirements

The property chooses to reconstruct into Fangcang shelter hospitals should be single-floor or multi-floors. The standard of construction structure, fire resistance level, fire compartment, firefighting facilities and fire lanes should comply and follow the existing relevant regulations.

The location of Fangcang shelter hospitals should be away from urban, crowded zone such as CBD, schools or residential areas, as well as factories contain flammable and combustible materials and hazardous chemicals. Signs and indication should be placed in areas beside the Fangcang shelter hospital. A green belt which should be no less than 20m and placed between hospitals and surrounding buildings. When actual condition is not in favor in building of green belt, the distance between hospitals and surrounding buildings should be no less than 30m.

There should be a parking lot and turning spaces for ambulances located in the entrance of the reconstructed property. The space should be sufficient enough for immediately evacuation, where ambulance can approach the outside traffic easily, but also directly connect to the internal control center. There should be extra rooms for medical supporting equipment, barrier free facilities, and logistics services in the parking lot. Moreover, the parking lot should be also provided with temporary tent, mobile medical facilities such as CT room and examination room, and living engagement facilities such as mobile toilet, wash room and bathroom. The interior space of the property should be easily separated into different zone using belts. Public venues (including exhibition centers, indoor stadium, vacant factory, and multi-function sports hall) with good firefighting facilities, is preferred.

3.2.2 Reconstruction Structure

If the construction design of Fangcang shelter hospitals involves changing the original loadbearing of the buildings, there should be a further evaluation and examination carried out by structural engineers. Precautions should be taken according to the actual situation and loadbearing adjustment made. Cautions should be taken on the following aspects:

(1) When transferring heavy weight medical equipment, evaluation on the layout plan and construction drawings should be carried out based on the loadbearing information. Precautions should be made according to actual condition and site evaluation results (the weight of the equipment should be less than the loadbearing, increase the loadbearing accordingly).

(2) When setting up the separation belt in between different areas, re-examination should be carried out based on the layout drawings and original loadbearing of the building. Precaution should be taken accordingly, for example using a lighter material for the separation belts.

(3) When transferring a heavy mobile facilities and equipment, reexamination should be carried out based on the actual weight of the equipment and its mobile path. Precaution should be taken accordingly.

(4) The separation belts should be installed safely and stably, and tightly linked to each other.

3.2.3 Firefight Facilities Requirement

(1) All existing firefighting equipment should be accessible and be able to use normally. The lightning system for evacuation should be in a good condition. Indication should be placed on the floor for clearly guiding the staff and patients to evacuate the building. Existed emergency exits should fulfill the relevant firefighting regulations and standards.

(2) Type A portable ammonium phosphate dry powder extinguisher should be equipped in the building. Severe hazardous area should be equipped at least with 3A level extinguisher, which the maximum area covered should be $50m^2$. The powder used should be ammonium phosphate MF/ABC5. The extinguishers in the building should be secured and placed according to relevant regulation.

(3) Fire extinguisher should be equipped in valuable equipment storage room, treatment rooms and information system control room.

(4) When domestic water supply system is installed under the condition where there is no fire hose reel, there should be an extra hose reel equipped. In the building (a portable fire hydrant is also applicable). The placement of such facilities should ensure the ratio of number of hydrant and surface area at 1∶1.

(5) Every medical staff should be provided with filtering respiration protective equipment. Its placement should be clearly seen and easily accessed.

(6) A small-scale firefighting station should be set beside the nurse duty room, equipped with a portable high-pressure water mist with a capacity of 100ml.

(7) If condition allows, all fire alarm and firefighting linkage control system should be ensured to work reliably.

3.2.4 Requirements on the Construction Site

(1) The designing process, purchasing process, construction and acceptance process should be carried simultaneously. There should be a close working relationship established between design team and construction team, which allow them to corporate with each other and deliver an outstanding service and performance.

(2) Different teams should be assigned to different zones and divisions, which take a modular approach, and deliver a standardized work performance. Duplicated work should be avoided, and reasonable time gap should be provided in between different construction tasks.

(3) Separation walls should be built based on the layout plans and construction drawings. Material used for the walls should be light and incombustible, which the flammability should be low and no less than level B1. Acceptance check should be carried out areas by areas in a timely manner. Examination and test should be presented on the rigidity, strength, stability and leakproofness of the separation walls.

(4) There should be extra precaution measures taken to increase the stability and strength of the walls, on supporting pipes and inside the separation walls. Walls and ceilings using light materials should be equipped with extra anti-cracking techniques.

(5) Regular examination should be carried out on ventilation system, power system and other related installed systems, in order to ensure all systems installed fulfilled the instructions on designed plans

and followed the guidance on relevant regulations and standards.

(6) Precaution measures should be taken to ensure a healthy, clean working environment for staff on-site; extra care should be provided to decrease the risk of infection. There should be an infrared detector in every entrance and exit, which supervises by staff who responsible for manual testing on staff's body temperature. Randomized check the staff's body temperature in every 4 hours.

(7) The toilets and offices for staff should be disinfected every 6 hours. Keep the working sites clean and maintain the room ventilation in a good condition.

(8) Smoking is strictly prohibited in working sites. Extra precaution should be made to emphasize the firefighting management. Decrease the use of open flame in the working site if it is possible. Installed extinguishers and small-scale firefighting station in the building according to the regulation.

(9) A spare dual power system should be installed in the building. Evert area should be equipped with an electric leakage protection device, to ensure the safety of the working site.

3.3 Reconstruction of Water Supply and Drainage

3.3.1 Construction Design Principle

The construction should be carried out based on the construction drawings and other relevant information drawings provided by a construction company. Reconstruction should strictly comply and follow the relevant regulations and standards.

3.3.2 Water Supply System

Water supply system in the Fangcang shelter hospitals should be reconstructed in the base of existing water supplying pipe in the property. An extra pressure-reducing and anti-backflow device should be installed in the entrance of water pipe in order to prevent the polluted water flow backward. Alternatively, a break tank could be used for water supplying. Water pressure should be maintained in at least 0.25MPa. Extra ports for domestic water pressure and chlorination should be reserved. Flushing and disinfection devices should be equipped in the parking lots for ambulances. For interior water supply pipes, S3.2, PPR should be used with hot-melt conjunction. For interior hot water supply pipe, S2.5, PPT should be used with hot-melt conjunction. When water pipe with DN<50mm on the pipe, copper stop valve should be used; while DN>50mm, high-temperature resistance stop valve should be installed on hot water supply pipes. The copper-core valve with ductile cast iron shell should be installed on pressure drainage pipe, with a nominal pressure of 100MPa.

3.3.3 Water Heating System

Electric water heating system should be implemented in the bathroom, with device to ensure ground protection in order to prevent burning, high pressure and overheated. Extra precaution measure should be taken to prevent situation including electric leakage, and automatic power cutting.

3.3.4 Water Boiling System

Each ward area should be complemented with a boiled water supply station, with provision of drinking water and boiled water adequately in a timely manner. The drinking water quality should be in adherence to the local *Sanitary Standard for Drinking Water*. Alternatively, if water boiling system is not feasible, it can be replaced with water dispenser.

3.3.5 Drainage System

Proper disinfection must be applied on disposed feces, vomitus, sewage, and medical liquid waste. Solid medical waste and chemicals cannot be directly disposed into sewer before being disinfected. It is strictly prohibited to dispose medical wastes and sewage from polluted area without disinfection.

Proper disinfection must be applied on domestic sewage from temporary toilets. Sewage must be collected centrally, and the disinfection procedure must comply with relevant domestic disposal regulation and standards. The water quality of disinfected sewage must meet the standard of local *Standard of Disposed Water Pollutants of Medical Institutions*. The disinfection procedure is as below:

(1) After the use of temporary toilets, immediately put an appropriate number of disinfectant tablets (peracetic acid, sodium hypochlorite, or bleaching powder) into the toilets. The sewage should be collected centrally by local environment department, which then transported to sewage disposal station and being disinfected. It is

prohibited to directly dispose the polluted sewage into urban drainage pipe network.

(2) The sewage from ward areas should be collected centrally, which then disposed into nearby septic tank. The liquid medical waste shall be distributed into existing drainage inspection tank. Liquid medical waste and sewage from ward areas should be disposed through separate drainage pipe, into separate septic tanks.

(3) Sewage from Fangcang shelter hospitals should undergo disinfection procedure twice before being distributed into urban drainage pipe. For property equipped with three-compartment septic tank: put appropriate amount of disinfection tablets into one of the compartments, which then distribute outward after at least 15 hours; disinfect the sewage for the second time at the entrance of the urban drainage pipe. The condition of the sewage should meet the requirement and standards set by local environment department.

(4) The duration of the exposure of sewage to liquid chloride, chlorine dioxide, sodium hypochlorite, bleaching powder, or calcium hypochlorite should be at least 15 hours, with residual chlorine more than 6.5mg/L, fecal coliform smaller than 100/L and choline dosage of 50mg/L. If the required duration of exposure is not feasible, the dosage of residual chlorine and available chlorine should be increased.

(5) For property without any sewage disposal or treatment facilities, temporary sewage collection tank or mobile septic tank should be installed, which allow hospital to efficiently dispose any polluted liquid.

(6) Ventilation pipes installed above the ceiling should be equipped with high-efficiency filter or UV disinfection device. The

equipment should be provided by mobile toilets suppliers.

(7) Sewage from vehicle wash station and disinfected wastewater should be disposed though drainage pipe. Sewage effluent should be sealed with proper facilities. Moveable mechanical piston is prohibited in sealing the effluent. The water depth should be no less than 50mm.

3.3.6 Installation of Drainage System

For Fangcang shelter hospitals drainage pipes, UPVs should be used, with plastic cement. Bathroom should be equipped with straight floor or grid-type floor drain. Water storage space should be reserved below the drain, with depth of no less than 50mm. When connecting sanitary facilities, which has no water storage space, with domestic drainage pipe or other pipe which may produce poisoned gas, there should be water storage space below the sewage effluent. The depth for water storage should be no less than 50mm. Drainage pipes shall be installed following the instruction below, otherwise comply with specification shown in construction drawings.

Standard slopes for installation of drainage pipes of Fangcang shelter hospital

Pipe size	*DN*75	*DN*100	*DN*150	*DN*200
Standard slopes for installation of sewage and waste water pipes	0.025	0.020	0.02	0.01

45° tee, 45° cross, 90° lateral tee, or 90° lateral cross must be used between horizontal pipes and between horizontal pipes and vertical pipes of drainage pipelines.

3.3.7 Sanitary Facilities

Automatic or touchless faucet sinks should be installed in every water-use site, and sanitary facilities with sterility standard, or facilities which used to epidemic prevention and control. Measures should be taken in order to prevent the splash of water or dirt. The following water-use site should be equipped with automatic or touchless faucet sinks:

(1) Basins for medical staff: including sinks and test tanks in bacteriological laboratory.

(2) Public toilets: Urinal and pedestal pan should be provided with automatic sensor faucets, while squatting pot should be provided with foot-pedal faucet or automatic sensor faucets.

(3) Basins for hand washing: Touchless faucet or foot-pedal faucet should be used.

3.4 Reconstruction of Ventilation and Air Conditioning Systems

3.4.1 Necessity of Ventilation System Installation

The existing ventilation and air condition systems installed in the properties before reconstruction into Fangcang Shelter Hospitals are mostly positive-pressure systems that can not meet the medical institution requirements and standards. The principle of reconstruction of ventilation and air condition system is to design and install additional facilities and devices to existing systems and to change the working mechanism of the systems, which allows different areas to have different room pressure. Always ensure there is the highest room pressure in clean areas, while lowest in the contaminated areas.

3.4.2 Construction Model

In order to accelerate the reconstruction and to fulfill the design requirements, the integrative reconstruction approach "Design-Procurement-Construction" ("EPC") can be adopted. While the engineering and construction contractor carry out the detailed engineering design of the project and procure all the equipment and materials necessary, in the meantime they should construct to deliver a functioning facility or asset into the existing property. There should be a close working relationship and efficient communication among design team, procurement team, and construction team in the designing stage: site investigation should be carried out, meanwhile all the functional facilities and materials should be prepared, with a detail working schedule formulate by construction team. When entering the construction stage, designers should carry out in-site services, in order to adjust designing plan according to actual condition.

3.4.3 Ventilation System Design Key Points

(1) Mechanical air supply and air exhaust systems should be installed in contaminated and semi-contaminated areas. Air should go through a high efficiency filter before exhausting. The filter should be placed in the entrance of exhaust blowers. Mechanical air supply and air exhaust systems can be installed in clean areas, alternatively, open air spaces for natural ventilation.

(2) Central air condition system for contaminated and semi-contaminated area should be equipped with air cleaning and disinfection devices. When it is feasible, higher-efficiency filter can be installed in the air condition machine. UV disinfection lamps can be installed near the return air filter and surface air coolers.

(3) Temporary air ventilation and exhaust facilities should be installed in the existing property based on construction drawings. The air flow direction should be from medical staff working areas to ward areas. There should be no dead corner left where air cannot be circulated properly.

(4) When existing air conditioning (AC) and ventilation systems are usable, they should be reset to DC air supply and exhaust system. Return air valves should be turned off, while air damper should be opened in full set. Full fresh air should be circulated into interior space, with rate of exhausting larger than rate of supplying. IF existing AC and ventilation system is not feasible, or there is no ventilation system installed in existing property, there should be additional functional facilities equipped during the reconstruction. The cabinet fan with an appropriate blowing rage and a suitable wind pressure should be installed to complement the existing systems. The height of the cabinet should be no taller than 2m, while equipped with precaution controlling measure. The AC and ventilation system should be working 24/7.

(5) The exhaust air rate should be set to 150m^3/h per person.

(6) When medical staff passing through clean areas and contaminated areas: the air supplying rate in the first dressing room should be no less than 30 times/h. D300 ventilation pipe should be

used in between dressing rooms and buffer rooms. When medical staff members return to clean zones from contaminated zones: The air exhausting rate in the buffer room and in room for removing PPE should be no less than 30 times/h. D300 ventilation pipe should be used in between dressing rooms and buffer rooms. The air direction of air flow should be from clean zones to contaminated zones.

(7) There should be functional facilities of sterilization and disinfection equipped in every isolated ward area. Additional heating system should be also installed based on actual condition.

(8) Emergency temporary dry toilets should be provided in isolated ward areas. Additional air exhaustion system should be installed in toilets for medical staff. The exhaustion rate should be no less than 12 times/h. High-efficiency filter should be equipped in the entrance of exhaustion outlet.

(9) The construction of ventilation and exhaustion system should be adjusted according to the actual situation; always ensure the fresh air circulating from outside. The outside environment near fresh air supply outlet should be clean and free from contamination. Outdoor exhaustion outlet should be installed in high altitude. The position should be above any supplying outlets no less than 20m, while the horizontal separation distance between exhaustion and supplying outlet should be no less than 6m.

3.5 Reconstruction of Electrical and Intelligence System

(1) Ventilation system should be centralized controlled in

nursing station (on-call room). The relevant equipment should be supplied in a full set.

(2) Light intensity should be adjusted in the reconstructed property in order to reduce glare impacts. Additional lamps can be installed on the wall of the open spaces; alternatively, upright light can be installed on the floor. Non-transparent covers should be provided on the lamps or using indirectly illumination facilities.

(3) There should be sufficient wireless network coverage, ensuring the accessibility of 4G and 5G network. When it is feasible, there should be provision of AP and WiFi network.

(4) Additional lamps, electricity wires and LV lines should be covered in metallic conduits or metal trunkings. The position of such conduits and trunkings should be set away from corridors. Precaution measure should be taken in order to prevention any incontinence caused by them.

(5) Sockets for UV sterilizer and air fresher should be installed in bathroom and buffer rooms. Power should be supplied through reserved electric circuits. The sockets for UV sterilizers should be indicated clearly and separated from those for lightning.

(6) Medical equipment room, bathroom, and any functional room where man can bath, should be equipped with supplementary equipotential bonding facilities.

(7) Alarm bottom should be positioned in nursing station (duty room). The signal should be directed to security staff station.

(8) Closed-circuit television (CCTV) cameras should be installed in medical staff areas and ward areas.

(9) Projector and screen should be provided in each resting and

entertainment areas, for social engagement need.

(10) When it is feasible, power sockets should be installed on the separation belt beside the ward beds. Insulation inspection should be carried out at least once before conducting electricity through wires. The insulation resistance testing voltage and insulation resistance between LV/ELV power circuit and lines in the grounds should be no less than 0.5Ω.

3.6 Ward Setting

Negative pressure in the ward area should be maintained by the ventilation system. The exhaustion rate should be set at no less than $200m^3$/hxp. Air should go through a low-efficiency filter (G4), followed by medium-efficiency filter (F8) and a high-efficiency filter (H11) before exhausting though a vertical pipe positioned at height. The outside door should be frequently opened, in order to supplement air using negative pressure.

The suppliers should deliver the ventilation and exhaustion pipes which equipped with fabric air duct. The outdoor exhaustion outlet should be positioned no lower than eave. A rodent-resistant net should be installed in the entrance of exhaustion outlet. The gap between each ventilation machine and doors, should be sealed properly.

Electrical air cleanser and fresher should be placed in the main corridors in ward areas and medical staff working areas.

3.6.1 Buffer Room Setting

Ventilation and exhaustion system should be installed in buffer room. The air supplying rate in the first-dressing room should be 30 times/h. A gate should be placed in between first-dressing and second-dressing room, so as to convey air from first-dressing to second-dressing room. Ventilation and exhaustion system should be installed in bathrooms and PPE doffing rooms.

High-efficiency filter and should be installed in the ventilation system. The size of the pipe should be D110, which is integrated by metal bellows and UPVC pipes.

3.6.2 Other Configurations

Temporary bathroom and shower places placed outside the Fangcang shelter hospitals should be provided with ventilation and exhaustion systems, which are equipped with high-efficiency filters. The size of the pipe should be D110, which is integrated by metal bellows and UPVC pipes. The exhaustion rate should be set at 8 times/h.

3.7 The Setting-up of Hardware and Software

3.7.1 Hardware Setting

Interior setting: the hardware requirements inside the Fangcang shelter hospitals would be mainly ward beds. It is highly suggested to use dorm-use or army-use bunk bed or foldable bed. Disinfection procedure can be applied, which can be also separated by physical belts made in eco-friendly materials. It is also necessary to have sufficient number of quilts, mats, and other bedding. If interior temperature is low, electrical blankets and hot pads should be also provided to medical staff and patients. If it is feasible, 5G network should cover the hospitals to provide convenience in telecommunication.

Exterior setting: High-end medical technology equipment should be provided, which include vans for imaging examination, testing, rescue work and P3 laboratory work.

3.7.2 Human Resource Management and Software Setting

The ration between patient number and number of medical staff, cleaning staff and food delivery staff should be reasonable and sufficient. Taken Zall (Wuhan Saloon) Fangcang shelter hospital as an example, in total there were 1169 medical staff, made up by 22 medical professional teams, accompanied by 15 nurse teams and 1 radiologic technologist team nationwide. Hierarchical management was adopted in managing these medical staff members, which were divided into 4 levels: director of medical affairs, director of hall, director of shelter, and doctors. There were in total 4 shifts that changes every 6 hours per day, namely A, P, N1, and N2. Each medical staff member had one day-off after working for 2 days. Every shift should be made up by 1 doctor and 5 nurses, which were responsible for taking care approximately 100 patients.

Apart from medical staff, there should be also staff members responsible for food delivery, security, cleaning, disinfection, psychological intervention, and heat and lightning maintenance. In Zall (Wuhan Saloon) Fangcang Shelter Hospital, there were approximately 100 people in logistic support team. Within the team, there were 54 cleaning staff members (3 shifts) who were responsible for disposing domestic and medical wastes, which weighted 600–700kg per day.

Chapter 4 Fangcang Shelter Hospitals Operation Model

4.1 General Rules and Principles

4.1.1 Management Principles

Targeted admission, centralized isolation, division management by units, standardized treatment procedure, and "two-way" referral.

4.1.2 Operation Objectives

Treat mild COVID-19 patients from community in an isolated environment, while control sources of infection to avoid cross-infection. In the meantime, provide various education program of COVID-19, as well as psychological consulting session, in order to maintain the mental stability of the patients. Frequent monitoring and treatment will be provided to patients in a timely manner, in order to prevent the patients' conditions from getting worse and decreasing mortality.

4.1.3 Organizational Structure

The Fangcang Shelter Hospitals is controlled and organized under a unified epidemic prevention and control center. The president of the hospital is fully responsible for its operation and management, and the vice-president is responsible for coordination and corporation. Staff are being divided by different teams and assigned with specific jobs and tasks.

(1) Integrated information and data control team (with assigned team leader): Preparation of operation schemes: determination of workflow; overall human resources management on division of labor and coordination; collecting, analyzing and reporting of various useful information; corporation and coordination on transferring; scheduling of administration and induction of medical and logistic staff; and finally but not least, coordination of operation issues.

(2) Medical operation team (with assigned team leader): Including medical groups and nursing groups. The medical groups are responsible for preparing health-care plans, developing relevant core systems and process, summarizing the information of medical staff profile and scheduling for them. The nursing groups are responsible for arranging nursing workflow and SOPs, including preparing nursing plans and processes, summarizing the information of nurses' profiles, and scheduling for them.

(3) Epidemic prevention and control team (with assigned team leader): Responsible for preparation and implementation of prevention and control system, provision of relevant training program, and nosocomial infection surveillance and monitoring during operation.

(4) Logistics supporting team (with assigned team leader): Responsible for efficiently allocating the medial and living resources and supplies, including providing medical equipment and facilities, preparation of medicines, environmental sanitation, disposal of medical waste, sewage discharge etc.

(5) Moreover, a functional management board can be established for each hospital. The size and function can be determined based on the actual situation and circumstances.

4.1.4 Working Mechanism

Each team shall have a clear division of function, which should be suitable for its own workflow and responsibility. Assign members into different shifts and plan the schedule according to the assigned shifts, then implement the system comprehensively. Besides, each team leader shall be responsible for the internal and external communication, coordination, information reporting and emergency situation to address the various problems arise from corporation. The scheduling system based on different time periods, shift change management system, and summary and report systems shall be implemented.

4.1.5 Work Discipline

(1) Be subject to overall command, have clear division of function, act proactively, and support each other.

(2) Ensure unobstructed communization and keep the phone available 24/7.

(3) Be punctual. Please report in advance if there is a reasonable excuse to do so.

(4) Do not disclose any confidential information in any inappropriate circumstances.

4.2 Admission Criteria

Taking account of the actual situation, the health condition of patients being admitted in Fangcang Shelter Hospitals should meet the following criteria:

(1) Present mild symptoms (mild clinical signs and symptoms, no evidence of pneumonia on imaging tests) or common symptoms (fever, respiratory symptoms or other symptoms, evidence of pneumonia on imaging tests).

(2) No clear functional disability.

(3) No evidence of severe chronic diseases (CHD), malignant tumors, structural lung diseases, pulmonary heart diseases and immunosuppression.

(4) No evidence on mental illness history.

(5) Do not have blood oxygen saturation (SpO_2) $> 93\%$ over a blood testing on finger, and a respiratory rate < 30bpm when resting.

(6) Any other medical history and situation should be clearly specified and explained.

4.3 Admission Procedure and Policy

(1) Before 10 a.m. every day, the head of the nursing group (head nurse) in each admission area of patients reports the number of patients that can be transferred to the director of the hospital

office according to the status of available beds. Then the director of the hospital office contacts with the vice president to determine the number of patients to be received on the same day and reports to the Epidemic Prevention and Control Command Center.

(2) According to the available ward beds capacity and admission capacity provided by Epidemic Prevention and Control Command Center, it should then confirm and provide the information of the number of patients, along with their profiles, to Fangcang shelter hospitals.

(3) Each Fangcang shelter hospital should establish a panel to review the status of the patients according to the admission standards, which helps in determining the list of patients whom to be received, as well as bed number assigned to them. There should be a proof certificate of transferring to every patient, which then report to the Command Center.

(4) Epidemic Prevention and Control Command Center is responsible for printing the information profile of each patients along with their Patient ID out, and then deliver it to the patients alongside with the Transfer Certificate.

(5) Epidemic Prevention and Control Command Center should arrange the transferring of the patients, coordinating ambulance dispatch and entourages on board, as well as managing the information of the ambulances. The patients' ID and their Transfer Certificate should be placed on the ambulance(Fig. 4-1).

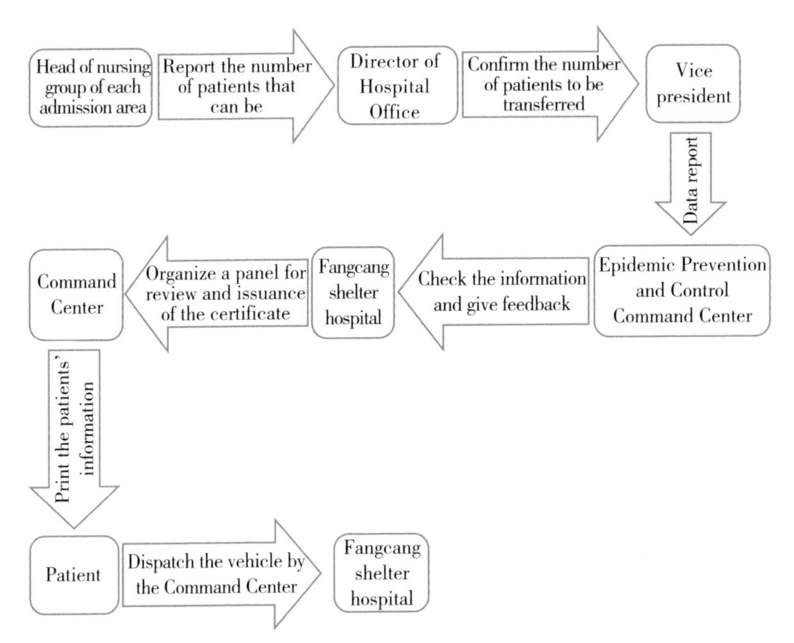

Fig.4-1 Admission procedure chart of Fangcang shelter hospital

4.4 Pre-examination and Triage

Fangcang shelter hospitals arrange the medical staff to carry out pre-examination and triage for patients. After pre-examination and relevant medical tests, the medical staff should give guidance to the patients who meet the admission criteria in hospitalization in a timely manner. The principle of "admission first, referral later" shall be followed when handle patients who do not meet the admission criteria while still presenting severe symptoms. To ensure medical safety, patients shall be organized in the observation areas for severe cases first, then provided with timely treatment as well as frequent monitoring on their health condition. Medical staff should contact designated hospitals and arrange the referral immediately.

4.5 Daily Health Check on Patients

Closely monitor the vital signs and oxygen saturation as per the following procedure:

(1) Measure and record body temperature 4 times a day at 8 a.m., 12 a.m., 4 p.m. and 8 p.m., respectively;

(2) Record respiratory rate (RR) two times a day, at 8 a.m. and 8 p.m., respectively;

(3) Measure heart rate (HR) and finger blood oxygen saturation two times a day at 8a.m. and 8p.m. respectively; Patients who have shown an unstable result may be applied with a finger clip oximeter to detect the transcutaneous oxygen saturation; frequent monitoring on oxygen saturation is required for patients with the severe disease until the condition is relieved. Alternatively, patients should be transferred to designated hospital.

(4) Whether to conduct laboratory tests or imaging tests is decided by the responsible physician according to the patient's condition.

(5) Whether to perform special examinations is decided by the responsible physician according to the patient's condition.

4.6 Intensive Care on Severe Patients

Severe patients refer to those who have been seriously ill at admission and the mild patients who have an exacerbation of the condition during hospitalization. In each ward area, there should be separate spaces for severe patents, which allow medical staff to closely monitor on them and provide timely treatment. The spaces shall be equipped with oxygen cylinders, rescue ambulance, rescue medicine, simple respirators, and monitoring and rescue equipment, as well as non-invasive ventilator and transfer wagon (where possible). A specially assigned medical staff should be responsible for the area, and allocation of the medical staff shall be strengthened and prioritized.

4.6.1 Indications for Initiating Consultation and Transferring to the Observation Area

(1) The subjective symptoms are not alleviated or even worsen despite treatment.

(2) Body temperature becomes higher than 38°C despite drinking of warm boiled water and physical cooling

(3) RR $\geqslant$ 30 bpm, which is not relieved after oxygen inhalation;

(4) Finger blood oxygen saturation $\leqslant$ 93% ;

(5) HR $\geqslant$ 100bpm and BP $\geqslant$ 140/90mmHg despite oxygen inhalation and antipyretic treatment.

4.6.2 Rescue Procedure for Severe Patients

Transfer the patients to the observation and treatment areas for severe cases with wheelchairs or wagons. Evaluate the state of illness, open the intravenous channels, and implement treatment. Provide life support and conduct monitoring. Apply for transferring the patients to a designated hospital. Record the medical conditions and report to higher-level medical institution.

4.6.3 Transfer Standards for severe patients

In principle, patients who show one of the following symptoms are considered to be transferred:

(1) Respiratory distress, RR>30bpm;

(2) Resting oxygen saturation <93% ;

(3) Arterial partial pressure of oxygen (PaO_2) / fraction of inspired oxygen (FiO_2) <300mmHg (1mmHg = 0.133kPa);

(4) The lesion shows a significant progression of >50 % within 24 to 48 hours as shown on lung imaging;

(5) With accompanying severe chronic diseases including hypertension, diabetes, coronary heart diseases, malignant tumor, structural lung diseases, pulmonary heart disease, and immunosuppression.

(6) Other reasonable specific reasons.

4.6.4 Transfer Procedure for Severe Patients

Patients in Fangcang shelter hospitals with unstable health condition or meeting the transfer criteria should be first consulted by doctors in specific area. After consultation, they can be transferred following the following procedure:

(1) Senior doctors will provide consultation to the patients after examination and triage, together with the responsible physician who escort the patients.

(2) Patients who are confirmed as severe case after consultation shall be reported to the Command Center immediately and transferred to the designated hospitals for further treatment;

(3) The transfer register sheet is filled up, and further induction from Command Center is waited;

(4) Coordinate to complete the handover process based on the instruction given; arrange medical staff to escort the transfer wagon; and keep the medical record of the patients updated.

4.7 Discharge Criteria and Procedure

4.7.1 Discharge Criteria

Patients can be discharged once the health conditions have met the following indication:

(1) Body temperature being stable over 3 days;

(2) Significant improvement of respiratory symptoms;

(3) Obvious absorption of inflammation as shown on lung imaging;

(4) Results of two consecutive (with at least 24 hours interval) NAT of respiratory pathogen being positive.

Patients showing the above indications can be discharged from the Fangcang Shelter Hospitals after getting consent from both specialist and head doctor.

4.7.2 Discharge Procedure

(1) The senior doctors will provide consultation to the patients after relevant examinations done by the responsible physician.

(2) Patients who meet the discharge criteria should be reported to the Command Center in a timely manner.

(3) Complete the discharge sheets and wait for the further instruction.

(4) Coordinate to complete the handover process based on the instruction given; arrange medical staff to escort the discharge; and keep the medical record of the patients updated.

(5) A clear explanation must be given to the discharged patients on the necessary of home-quarantine: self-isolation in a single room while wearing a mask; avoid going out; and body temperature must be measured every day during the 14-day home quarantine. If home-quarantine is not feasible, the patients should follow the instruction and centralized quarantine arranged by the Command Center. If the COVID-19 symptoms recur or worsen , the conditions must be reported to the community workers immediately. The patients should immediately seek medical intervention in a designated hospital.

4.8 Disinfection and Disposal Procedure

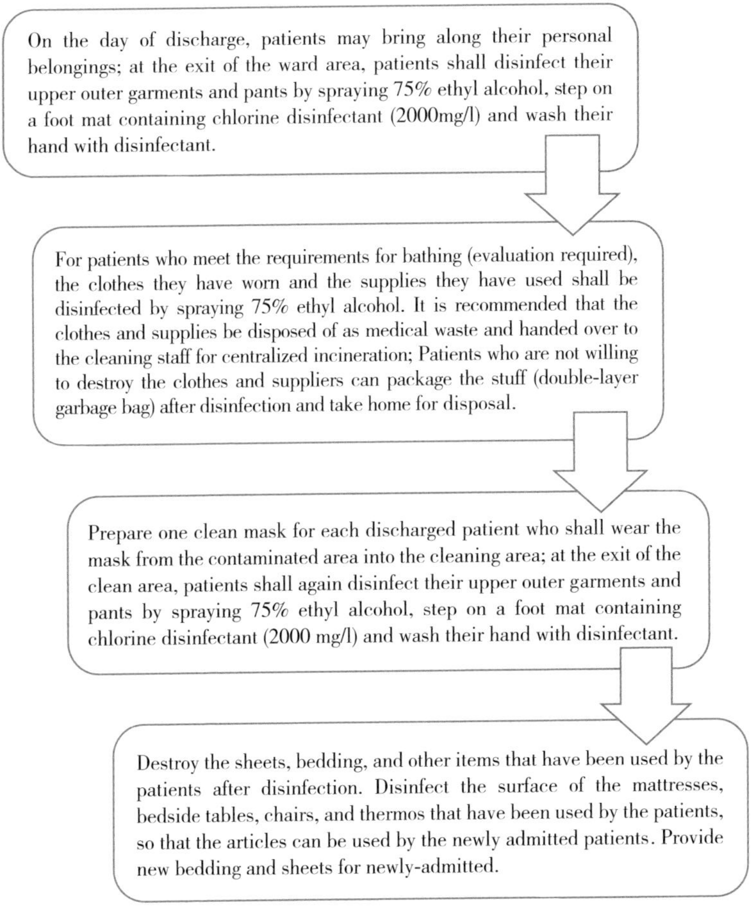

On the day of discharge, patients may bring along their personal belongings; at the exit of the ward area, patients shall disinfect their upper outer garments and pants by spraying 75% ethyl alcohol, step on a foot mat containing chlorine disinfectant (2000mg/l) and wash their hand with disinfectant.

↓

For patients who meet the requirements for bathing (evaluation required), the clothes they have worn and the supplies they have used shall be disinfected by spraying 75% ethyl alcohol. It is recommended that the clothes and supplies be disposed of as medical waste and handed over to the cleaning staff for centralized incineration; Patients who are not willing to destroy the clothes and suppliers can package the stuff (double-layer garbage bag) after disinfection and take home for disposal.

↓

Prepare one clean mask for each discharged patient who shall wear the mask from the contaminated area into the cleaning area; at the exit of the clean area, patients shall again disinfect their upper outer garments and pants by spraying 75% ethyl alcohol, step on a foot mat containing chlorine disinfectant (2000 mg/l) and wash their hand with disinfectant.

↓

Destroy the sheets, bedding, and other items that have been used by the patients after disinfection. Disinfect the surface of the mattresses, bedside tables, chairs, and thermos that have been used by the patients, so that the articles can be used by the newly admitted patients. Provide new bedding and sheets for newly-admitted.

Fig. 4-2 Disinfection for discharged patients of Fangcang shelter hospitals

Chapter 5 Fangcang Shelter Hospitals Logistics Support

To provide logistics support for Fangcang Shelter Hospitals, we formulate the following plans from the aspects of catering, accommodation, cleaning, and supplies, based on the actual logistic situation.

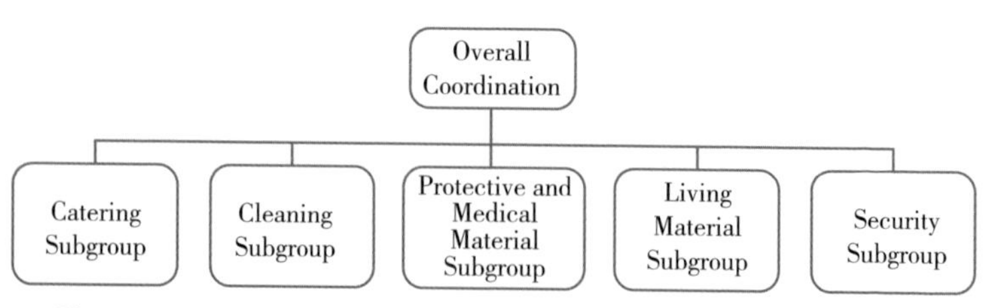

Fig. 5-1 Logistics support chart of Fangcang Shelter Hospitals

5.1 Materials and Supplies Support

5.1.1 Catering Support

5.1.1.1 Staffing

One manager in charge shall provide his/her contact information and coordinate his/her work with other departments.

5.1.1.2 Work Plan

(1) Count the number of patients and staff on a daily basis, prepare meals in advance, and ensure that they are sufficient.

(2) Distribute food according to the name list and preplanned timeline.

(3) Contact catering companies.

(4) Ensure cleanliness, safety, and hygiene of the diets.

(5) Ensure daily delivery of fresh dishes and their safety. Impose strict quantity and quality check.

(6) Suggested breakfast time is 7:00 to 8:00, lunch time is 11:30 to 12:30, and dinner time is 17:30 to 18:30.

5.1.1.3 Requirement

(1) Deliver the meals of patients and medical workers to the designated areas and notify them in WeChat groups or with other means of communication.

(2) Record the distribution of breakfast, lunch, and dinner to patient and staff.

5.1.2 Hygiene Support

5.1.2.1 Staffing

Two managers in charge shall provide their contact information and coordinate their work with other departments.

5.1.2.2 Work Plan

(1) Arrange cleaning worker of each area, including public areas, patient areas, clean areas, toilets/ wash rooms, and other locations.

(2) Assign cleaners based on demand. Dispose of garbage in time.

(3) Redeploy and reallocate staff when needed.

(4) Educate the patients to ensure their personal hygiene.

5.1.2.3 Requirement

(1) Clean up after breakfast, lunch, and dinner on a regular basis. Enforce rectification if the requirement is not met. Require the cleaning staff to sign in at the responsible area every hour.

(2) Garbage cleaning. Provide garbage bins such as on walkways, doorways, dining areas or each floor. And the cleaning staff should move the garbage to the garbage room in a timely manner.

5.1.3 Medical Supplies Support

5.1.3.1 Staffing

Two managers in charge shall provide their contact information and coordinate their work with other departments.

5.1.3.2 Work Plan

(1) Make lists and ensure the supplies of personal protective equipment according to the daily schedule of doctors and nurses.

(2) Register the usage of protective equipment according to the temporary work arrangement. All PPE shall be received within hours to avoid wasting.

(3) Keep records of the collection of protective clothing and ensure the record to be accurate and consistent.

(4) Set up 24-hour PPE distribution posts. Supplies such as oxygen bottles and medicines shall be provided to the patients when necessary.

5.1.3.3 Requirements

(1) When the inventory of PPEs is less than 100 sets (the specific

standard depends on the number of medical personnel in the shelter), report and coordinate immediately.

(2) At the distribution office, explain the correct use of PPEs and avoid cross-infection.

(3) At the distribution office, the recipients should identify themselves and sign their names before receiving PPEs.

5.1.4 Life Supply Support

5.1.4.1 Staffing

Two managers in charge provide their contact information and coordinate their work with other departments.

5.1.4.2 Work Plan

(1) Provide patients with sufficient daily necessities such as quilts, electric blankets, cups, pots, towels, etc.

(2) Coordinate the supply of water, electricity, and network service to ensure that patients have hot water and electric power available

(3) Arrange office supplies such as computers, stationery, tables and chairs in the working areas.

(4) Establish connection with the donation office to ensure that the donated materials are stored and distributed safely and make the best use of the donated items.

5.1.4.3 Requirements

(1) The registration list shall be completed according to the bed number of the patient when the supply is collected.

(2) Record the distribution of materials to ensure that the items are sufficient.

5.1.5 Management of Medicines

5.1.5.1 Formulate a Drug Catalog for a Fangcang Shelter Hospital

According to the characteristics of the patients admitted in the hospital, the relevant diagnosis and treatment plans and guidelines, and the opinions of front-line experts, the drugs needed for COVID-19 treatment in a Fangcang Shelter Hospital shal be evaluated. The catalog should be mainly composed of drugs for symptomatic treatment, prevention of complications, treatment of underlying diseases, and first aid. The final catalog of drugs is determined by the clinical pharmacist and purchasing department. In the later stage, according to the actual clinical situation, regularly adjust the quantity and variety of drugs.

The catalog includes the following medicine:

(1) Antiviral, antibacterial, analgesic-antipyretic, antitussive, antiasthmatic and expectorant, gastrointestinal drugs;

(2) Hypnotics and sedatives;

(3) Drugs for blood pressure or glucose reduction, lipid regulation and other chronic diseases;

(4) Chinese patent medicines or other traditional medicines clinically tested to be effective in the rehabilitation of COVID-19 patients;

(5) Drugs for first aid.

5.1.5.2 Establishment of Hospital Pharmacy

Take Zall (Jianghan Wuzhan) Square Cabin Hospital as an example, the hospital pharmacy covers an area of about $30m^2$and is set in a clean zone and divided into qualified area, unqualified area and secondary storage area. The hospital is equipped with basic facilities such as computers, printers, fire protection, and anti-theft to ensure that the storage conditions of medicines meet the requirements of relevant management regulations.

5.1.5.3 Procurement and Supply

Drug supply team is in charge of drug, who should formulate unified procurement and request plans. The pharmaceutical emergency leading team determines the list of medicines to be procured urgently in accordance with the established hospital pharmacy drug catalog.

Other than ensuring common medical supplies, the hospital should also focus on ensuring adequate supply of drugs related to COVID-19 treatment and store the drug in specific area. The procurement must be made from pharmaceutical companies with legal qualifications. The qualifications of relevant companies and business personnel must be recorded alongside the procurement for transparency and accountability. When there is a shortage of supplies, the hospital should actively negotiate and communicate with the providers, encouraging them to expand the supply chain or transfer the supplies from other regions. Meanwhile, if the procurement is difficult, the hospital should actively seek alternative supplies and materials and publish drug guidelines to the clinic.

5.1.5.4 Management of donated drugs

(1) Acceptance criteria for donated drugs

1) For those produced within China, the drug must be a product approved by the local drug regulatory authority with a valid approval number, and meet quality standards. The drug must be valid and

effective for at least 6 months.

2) For drugs produced outside China, the varieties shall be approved and registered by local regulatory authorities, included in the international general pharmacopoeia and legally produced and marketed in registered countries, and meet quality standard; the expiry date is more than 6 months.

(2) Standard process for accepting donated medicines

1) Formulate a list of donated medicine: Discuss and develop a list of medicines acceptable for donation and update them regularly. The medicines already in the catalogue can be accepted as donations directly.

2) Receiving process of donated drugs outside the catalog: After receiving the donated medicine, the drug supply group submits the product information and instructions provided by the donors to clinical pharmacy for clinical demonstration and evaluation or consults the clinical medical team and clinical experts to determine whether to accept the donation.

3) Quantity of medicines accepted for donation: The amount of donated drug is determined by pharmaceutical emergency leadership team based on the usage of drugs, the time and the stage of the epidemic.

5.2 Living and Social Engagement Support

The Fangcang Shelter Hospitals serve as a "special community" for patients with mild disease. In order to enrich the cultural life of patients and enhance their courage to overcome the virus, it is recommended to set up the following facilities in the Fangcang shelter hospitals:

(1) Reading corner: Each hospital is equipped with a bookshelf, where interesting books are placed for patients to read freely. See Fig. 5-2.

Fig. 5-2 Book corner of Zall (North Hankou) Fangcang Shelter Hospital

(2) Food corner: Each hospital can set up a caring corner with foods such as instant noodles, milk, and fruits. See Fig. 5-3.

Fig. 5-3 Food corner of Zall (Wuhan Salon) Fangcang Shelter Hospital

(3) Power-charging station for electronic devices: Each hospital should have one or more free charging stations for patients to charge their electronic devices. See Fig. 5-4.

Fig. 5-4 A free charging station in Zall (Wuhan Salon) Fangcang Shelter Hospital

(4) Entertainment corner: One or two TV sets are installed in the open space in each hospital. Such an entertainment corner can serve as an area for patients to watch various TV programs; in addition, the area can also be used for organizing square dance, game programs, poetry recitation, chorus, and other literary programs to please and encourage the patients. See Fig. 5-5 and Fig. 5-6.

Fig. 5-5 An entertainment corner in Zall (Wuhan Salon) Fangcang Shelter Hospital

Fig. 5-6 Square dancing at the entertainment corner of Zall (Wuhan Salon) Fangcang Shelter Hospital

(5) Psychological counseling: The sudden COVID-19 epidemic brought a series of stress disorders and anxiety to the patients. Thus, mental health counseling is required.

5.3 Safety Support

5.3.1 Medical Waste Management System in Fangcang Shelter Hospitals

(1) Accountabilities. Medical waste generated in the Fangcang Shelter Hospitals shall be properly managed by implementing the accountabilities of all staff members. The person in charge of each area is the first person responsible for medical waste management, and the operators who produced medical waste are the direct persons responsible for waste disposal. Efforts should be made to strengthen environmental sanitation and hygiene and dispose medical waste timely.

(2) Training system. All staff including physicians, nurses, technicians, administrators, and workers receive unified training from the Infection Control Team.

(3) Supervision system. The Infection Control Team is responsible for regular inspections and for problem identification, feedback, and rectification on the collection and disposal of medical waste in each area.

(4) Waste sorting. Waste produced by medical institutions or hospitals during the treatment of suspected or confirmed COVID-19 patients, including medical waste and garbage, should be collected,

classified, and disposed in the same way as medical waste.

(5) Establishing strict standards for packaging and containers. Warning signs must be attached to the surface of special packaging bags and sharps containers for medical waste. Before disposing medical waste, careful inspection should be carried out to ensure the container is not damaged or leaking. Containers which are foot-operated and with lids are preferred for waste collection. When the medical waste reaches 3/4 of the packaging bag or sharps container, it should be sealed effectively and tightly. Double-layer packaging bags should be used to contain medical waste, and gooseneck-type seals should be applied for layered sealing.

(6) Collecting waste safely. According to the type of medical waste, collect the waste in a timely manner, ensuring the safety of personnel and minimizing the risk of infection. When the surface of the packaging bag and sharps container is contaminated with infectious waste, a layer of packaging bag should be added. It is strictly forbidden to squeeze the bag when collecting disposable clothing, protective clothing and other items after use. Each packaging bag and sharps container should be attached with a label. The label should include the name of medical institution, department, date and type of waste, and mark specifically "Coronavirus Disease 2019" or "COVID-19" in short.

(7) Processing the waste from different zones. For those medical waste produced by suspected or confirmed COVID-19 patients at fever clinic and wards, the surface of the packaging bag should be sprayed uniformly and disinfected with 1000mg / L of chlorine-containing disinfectant, or a layer of medical waste packaging bag

should be added on the outside before leaving the contaminated zone. The medical waste generated in the clean zone should be disposed of as conventional medical waste.

(8) Handling pathogen sample. High-risk waste such as pathogen-containing specimens and preservation solutions in medical waste should be pressure-steam sterilized or chemically sterilized at the place of production and then collected and disposed of as infectious medical waste.

(9) Central collection and transferring of the medical waste.

1) Management of safe transportation. Before transporting medical waste, the personnel should check whether the label and the seal of the bag or sharps container meet the requirements. While transporting the waste, the personnel should prevent damage to the packaging bags and sharps containers filled with medical waste and avoid direct contact, leakage, or spread of medical waste. Clean and disinfect the transportation tools with 1000mg/L chlorine-containing disinfectant after the daily transportation. When the transportation tool is contaminated with infectious medical waste, it should be disinfected immediately.

2) Management of handover and storage. The temporary storage place for medical waste should be tightly closed and managed by staff. Avoid irrelevant personnel from coming into contact with medical waste. Medical waste should be stored in a separate and temporary area and handed over to the medical waste disposal unit for disposal as soon as possible. The floor of temporary storage places must be disinfected with 1000mg/L chlorine-containing disinfectant twice a day. The medical departments, transport personnel, temporary storage staff, and medical

waste disposal unit transfer personnel should check and register while handing over the waste and explain to each other that it originates from suspected or confirmed COVID-19 patients.

3) Transfer registration. Strictly implement the joint management of hazardous waste transfer and registry of medical waste. The registered content includes the source, type, weight or quantity of medical waste, handover time, final destination, and signature of the manager. It must be marked specifically as COVID-19, and the registration information must be kept 3 years. The medical waste disposal unit shall be notified in time for door-to-door collection or self-built medical waste disposal site, and the corresponding records shall be made. Health administrations and the hospitals should strengthen the communication with the ecological environment departments and medical waste disposal units to optimize the disposal of medical waste during the epidemic.

5.3.2 Nosocomial Infection Control

5.3.2.1 Objective

Hospital infection control is to reduce the risk of virus transmission in the hospital, standardize the practices of all staff (including medical staff) to avoid infection.

5.3.2.2 Teamwork

A nosocomial infection committee, composed of hospital president, vice-presidents in charge of medical affairs, nosocomial infection, nursing and logistics, ward head nurses, and ward administrative directors shall be established. The vice-president in charge of nosocomial infection leads the establishment of nosocomial infection control teams, which include ward head nurses, nosocomial infection physicians, nosocomial infection nurses, and logistics department liaisons.

5.3.2.3 Work plan

(1) Area division. Clearly define the contaminated area, semi-contaminated area, and clean area in the hospital, and post a notice at a conspicuous location in the hospital. A prominent warning sign shall be set at the junction of each district. Special entrances and exits must be arranged at the entrance and exit of the contaminated area to ensure compliance with hospital infection regulations.

(2) Staff training. All the staff members shall strictly abide by the system of personnel training before taking up positions. The training content is tailored for different personnel. All personnel who need to enter the contaminated area should be given key training; they should grasp the knowledge, methods, and skills of nosocomial infection control and raise the awareness of prevention and control. All the workers and personnel should be trained on nosocomial infection control and assist them to perform well in environmental cleaning and disinfection, patient transfer, and medical waste disposal.

(3) Protection of medical staff and workers. On the basis of standard prevention and control measures, the prevention and control of contact transmission and airborne transmission shall be strengthened. Select and use PPEs properly and implement hand hygiene strictly.

5.3.2.4 Prevention and control system

(1) Grading of PPEs

Grading of PPEs in Fangcang Shelter Hospitals

PPEs	Primary protection	Secondary protection	Tertiary protection
Hat	*	*	*
Isolation gown	*		
Protective suit		*	*
Disposable surgical mask	*		
Medical protective mask		*	*
Goggles/protective face shield		Alternative	Both
Gloves	*	*	*
Boot/protective shoe cover		*	*

(2) Protection levels in different areas

Protection levels in different areas

Working area / work content	Primary protection	Secondary protection	Tertiary protection
Contaminated area		*	
Semi-contaminated area		*	
Clean z area	*		
Specimen collection (specimens from the respiratory tract)			*
Specimen collection (specimens from the non-respiratory tract)		*	
Specimen delivery		*	
Disinfection (contaminated zone and semi-contaminated zone)		*	
Disinfection (clean zone)	*		

(Continued)

Working area / work content	Primary protection	Secondary protection	Tertiary protection
Patient escort and transfer		*	
Nucleic acid testing (NAT)			*
Laboratory test (specimens from the non-respiratory tract)		*	
Imaging examination		*	

(3) Donning/Doffing of PPEs

(4) Hygiene and disinfection of environments

In accordance with the *Technical Guidance on Disinfection in Medical Institutions*, the diagnosis and treatment environments (air, object surface, ground, etc.), medical equipment, and patient materials should be cleaned and disinfected strictly. Patients' respiratory secretions, excreta and vomit shall be disposed of properly. Sterilization and cleaning appliances in contaminated areas, semi-contaminated areas, and clean areas must be marked with different colors and must not be mixed.

(5) Patient management and education

Patient should be educated on personal protection and cough etiquette. Masks should be worn throughout the transfer or when receiving auxiliary inspections and imaging studies out of the Shelter Hospital.

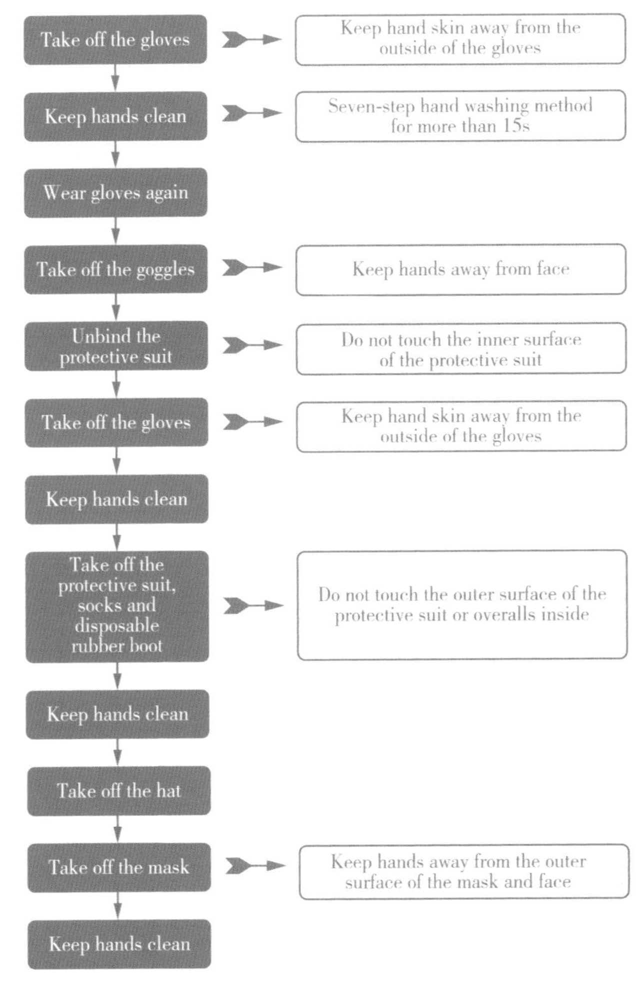

Fig.5-7 Standard operating procedure for removal of protective suits in the Fangcang shelter hospital

5.3.3 Security Support

5.3.3.1 Staffing

One manager in charge shall provide his/her contact information and coordinate his/her work with other departments.

5.3.3.2 Work plan

(1) Prohibit unrelated personal from entering and exiting the hospital at will.

(2) Prevent and stop riots and disputes.

(3) Coordinate and assist the transportation of supplies.

5.3.3.3 Requirement

(1) Assign at least two security guards at the entrance and exit of the hospital.

(2) 24-hour security patrol in the patient areas is required.

5.3.4 Emergency Plan for Fangcang Shelter Hospital

5.3.4.1 Emergency plan for unexpected disputes

(1) Objective: Prevent individuals inside the Shelter Hospital from harming others; prepare for any disputes or violence in the Shelter Hospital.

(2) Emergency procedures

5.3.4.2 Emergency plan for water or power failure

Work objectives: Plan should be established to properly respond to water and power failure, prevent and control water and power outage damage, keep normal diagnosis and treatment order, ensure the safety of patients and medical personnel, and maintain the safety and stability of the hospital. Emergency procedures include:

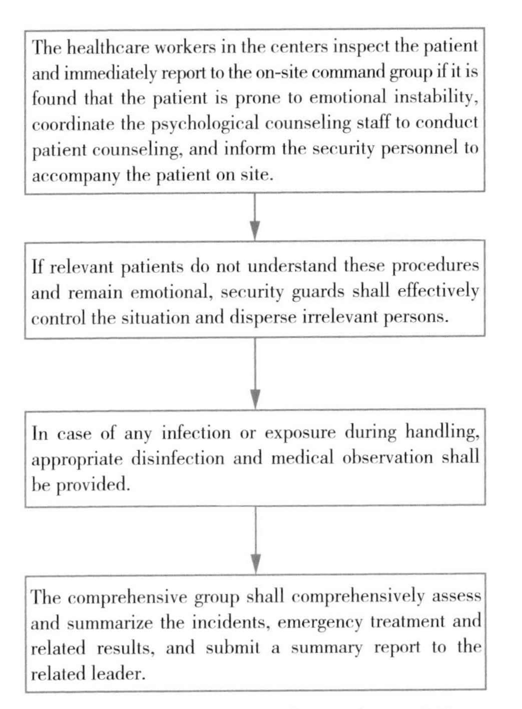

Fig. 5-8　Emergency Response Procedure of Fangcang Shelter Hospital

(1) Information report: When a water or power failure occurs, the first person to report the problem and each receiver must immediately report in accordance with the procedures and requirements of the emergency plan.

(2) Early handling: After an emergency occurs, each respondent must complete early information handling while completing the information report or initiate an on-site emergency plan according to their responsibilities and prescribed authorities and respond to control situations in a timely and effective manner.

(3) Emergency response: If the early response fails, a special

emergency plan for water and power outages should be started in time. The person in charge should direct the relevant departments to carry out the processing work. The medical treatment team leader should be responsible for directing emergency treatment. The medical staff should organize the rescue work in this area according to the on-site treatment plan for water and power failure of the department.

In addition, to ensure the emergency water and electricity supply:

(1) The members of the logistics working group should ensure that the communication is smooth at all times.

(2) Check the water supply pipelines, circuits, and valves on a daily basis, and deal with problems in a timely manner.

(3) Train relevant logistics staff members who are familiar with emergency water supply, pipeline layout, and emergency operation.

(4) The maintenance personnel of the water and electricity should be on duty for 24 hours.

5.3.4.3 Fire incident emergency plan

Work objective: To ensure the safety of personnel and property in the hospital area. In the event of a fire, effective measures should be taken to avoid or mitigate casualties and property losses as much as possible. Relevant personnel must master the escape skills. In case of an emergency, they should promptly evacuate to the nearest emergency evacuation site by using safe evacuation channels.

Emergency procedure:

(1) When a fire occurs, the person in charge on the spot should immediately organize the personnel in duty to use the nearest fire extinguishing equipment to extinguish the fire. Fire should be

extinguished according to the nature of the fire (for example, when an electrical fire occurs, the relevant power switch of the fault location should be cut off as soon as possible). Notify the security personnel to confirm the specific situation of the fire and dial 119 to call the police.

(2) Evacuating the crowd. Send evacuation instructions to the scene of the fire through emergency broadcasts. Medical staff on duty in each area should guide patients to evacuate from the fire scene in an orderly manner. The staff of the evacuation guidance group should have clear division of labor and be under unified command.

(3) Notifying the medical staff resting at the nearest hotel to treat the wounded in time at the scene. Contact other neighboring hospitals for treatment if necessary.

(4) Security staff should quickly arrive at the fire site to conduct on-site alert and maintain order.

(5) The members of the logistics team should register and keep the rescued and transferred materials and cooperate with relevant departments to clean up and register fire losses.

5.3.5 Operation and Maintenance of Ventilation System

(1) The operation scheme is developed according to the division of contaminated areas, semi-contaminated areas, clean areas, medical personnel passages, and patient passages. Open or close some new air valves, supply air valves, and exhaust valves. Adjust the angle of some air valves. Open or close air vents or windows. Open or close some ventilation and air conditioning equipment.

(2) Monitor the failure alarm signals of air blower and exhaust fan at all time to ensure its normal operation. Monitor the pressure

difference alarm of the air filters at all levels of the supply and exhaust system. Replace the blocked air filter in time to ensure the volume of the air supply and exhaust.

(3) The air handling unit and the new fan group should be checked regularly to keep it clean.

(4) Low-efficiency filter fans of the ventilation system should be cleaned every 2 days. The low-efficiency filters are recommended to be replaced every 1 to 2 months. Medium-efficiency filters should be checked weekly and replaced every 3 months. Sub-efficient filters should be replaced every year. Contaminated or blocked devices should be replaced in time. The terminal high-efficiency filters should be checked once a year and should be replaced when their resistance exceeds the designed initial resistance of 160Pa or if they have been used for more than 3 years.

(5) The medium-efficiency filters in the exhaust fan group are recommended to be replaced every year. Polluted or blocked medium-efficiency filters should be replaced in time.

(6) Check the return air filter regularly, and clean it once a week. If there is any special contamination, replace it in time, and wipe the inner surface of the air outlet with disinfectant.

(7) Assign special maintenance management personnel to maintain the fans following equipment instructions. Develop an operation manual for record.

(8) Operators who replace high-efficiency air filters on exhaust air must protect themselves carefully. The dismantled exhaust high-efficiency filters should be sterilized on site by professionals. The filters should be put into a safe container for sterilization and

sterilization, and be treated together with other medical waste.

5.4 Volunteer Service

As seen in previous emergency events, government should integrate social resources effectively when professional resources are not adequate and sufficient to respond to a public health emergency. People who register as volunteer must meet the relevant criteria and requirements. Most importantly, they should volunteer under their freewill and expect no economic return.

5.4.1 Criteria and Requirement

(1) Volunteers must strictly follow the staff arrangement and schedules set by the epidemic control teams when participating in volunteer events.

(2) Volunteers in Fangcang Shelter Hospitals must be in a good health condition and wear PPEs when participating in relevant work. They should receive adequate training before starting volunteer work. Volunteers should stay in their own working area. Cross-area service is strictly prohibited.

(3) People with a medical or psychological education background are preferred. Volunteers should also be familiar with epidemic control and prevention campaign, relevant policies and regulations, and knowledge and theories relating to psychological intervention.

(4) When offering services, volunteers should always apply the principle of "safety-first". Education and training on epidemic control

and prevention should be given regularly. Improper wearing of PPEs is prohibited. Volunteers who fail to meet the training expectations are not allowed to approach their work. Reasonable job division and schedule should be implemented. The number and shift time of volunteers should be closely monitored.

5.4.2 Job Division Scheme

Professional medical assistant: volunteers who enter high-risk contaminated area to assist medical staff in daily management of patients.

Medical supplies logistics: volunteers who deliver medical supplies including medicines, equipment, facilities, and PPEs to the assigned areas. These volunteers are also responsible for organizing and recording the medical supplies.

Living engagement support: volunteers who are responsible for providing social and living support for patients and medical staff. This includes transportation, diary and food supplies, and other living engagement supports. These volunteers are also responsible for the maintenance of the power, water, and heating supplies.

5.4.3 Job Design

(1) Professional medical assistant: These volunteers are responsible for maintaining the public order in Fangcang Shelter Hospitals, controlling the medical supplies in urgent need, and organizing the information of admitted patients. Take Zall (Wuhan Saloon) Fangcang Shelter Hospitals as an example, there were 54 volunteers who were responsible for cleaning and disinfection of

the equipment and facilities. They worked in three shifts per day, disinfecting and disposing the medical waste 7/24.

(2) Medical supplies logistics: Their main responsibilities are to transport, deliver, and organize various medical supplies including medicines, PPEs, respirators, surgery masks, and gloves.

(3) Living engagement support: The main responsibility is to ensure the essential living and social engagement activities of medical staff and patients. This includes the diary plan, transportation, and other engagement support related to culture and social engagement. Meantime, they also take the responsibility of maintaining and monitoring on the power, water and heat supplies in the hospitals, in order to respond to emergency events in a timely manner. Take Zall (Wuhan Saloon) Fangcang Shelter Hospitals as an example, there were approximately 100 volunteers participating in the reconstruction of the hospitals. They set up a reading corner, power-charging center, food corner, entertainment center, and other functional room. Moreover, they ensured the distribution of supplies donated by public community. They also monitored the power, water, and heating systems to prevent any malfunction and failure.

References

[1] Chen S, Zhang Z, Yang J, et al. Fangcang shelter hospitals: a novel concept for responding to public health emergencies [J]. The Lancet, 2020, 395: 1305-1314.

[2] Medical Administration and Hospital Authority of National Health Committee of the PRC, Medical Management and Service Guidance Center of National Health Committee of the PRC. Work Manual for Fangcang Shelter Hospital (3rd Edition) [S]. 2020.

[3] Department of Housing and Urban-Rural Development of Zhejiang Province. Technical Guidelines for Centralized Receiving and Treatment in Fangcang Shelter Hospital (Trial) [S]. 2020.

[4] Department of Housing and Urban-Rural Development of Hubei Province. Technical Requirements for Design and Reconstruction of Fangcang Shelter Hospital (Revised Edition) [S]. 2020.

[5] General Office of Ministry of Housing and Urban-Rural Development of the PRC, General Office of National Health Committee of the PRC. Guidelines on Design of Emergency Treatment Facilities in COVID-19 (Trial) [S]. 2020.

[6] Gu Ming, Hua Xiaoli, Chen Jun, Zeng Fang, Zhou Tao, Zhang Yu, Shi Chen. Pharmacy administration and pharmaceutical care practice in Jianghan Fangcang Shelter Hospital [J]. China Pharmacist, 2020, 23 (04): 702-706.

Acknowledgements

Jack Ma Foundation

Alibaba Foundation

China First Metallurgical Group Co., Ltd.

CITIC General Institute of Architectural Design and Research Co., Ltd.

Department of Housing and Urban-Rural Development of Hubei Province

Department of Housing and Urban-Rural Development of Zhejiang Province

Central-South Architectural Design Institute Co., Ltd.

Wuhan Real Estate Group Co., Ltd.

Wuhan Hanyang Municipal Construction Group Co., Ltd.

Hubei Yangtze River Industrial Investment Group Co., Ltd.

Wuhan Shooting School